PRACTICAL RETINAL OCT

PRACTICAL RETINAL OCT

Bruno Lumbroso MD
Director
Centro Oftalmologico Mediterraneo for Retinal Diseases
Rome, Italy

Marco Rispoli MD
Centro Oftalmologico Mediterraneo for Retinal Diseases
Rome, Italy

JAYPEE The Health Sciences Publishers

New Delhi | London | Philadelphia | Panama

Jaypee Brothers Medical Publishers (P) Ltd

Headquarters

Jaypee Brothers Medical Publishers (P) Ltd
4838/24, Ansari Road, Daryaganj
New Delhi 110 002, India
Phone: +91-11-43574357
Fax: +91-11-43574314
Email: jaypee@jaypeebrothers.com

Overseas Offices

J.P. Medical Ltd
83, Victoria Street, London
SW1H 0HW (UK)
Phone: +44 20 3170 8910
Fax: +44 (0)20 3008 6180
Email: info@jpmedpub.com

Jaypee-Highlights Medical Publishers Inc.
City of Knowledge, Bld. 237, Clayton
Panama City, Panama
Phone: +1 507-301-0496
Fax: +1 507-301-0499
Email: cservice@jphmedical.com

Jaypee Medical Inc
The Bourse
111 South Independence Mall East
Suite 835, Philadelphia, PA 19106, USA
Phone: +1 267-519-9789
Email: jpmed.us@gmail.com

Jaypee Brothers Medical Publishers (P) Ltd
17/1-B Babar Road, Block-B, Shaymali
Mohammadpur, Dhaka-1207
Bangladesh
Mobile: +08801912003485
Email: jaypeedhaka@gmail.com

Jaypee Brothers Medical Publishers (P) Ltd
Bhotahity, Kathmandu, Nepal
Phone: +977-9741283608
Email: kathmandu@jaypeebrothers.com

Website: www.jaypeebrothers.com
Website: www.jaypeedigital.com

© 2015, Jaypee Brothers Medical Publishers

The views and opinions expressed in this book are solely those of the original contributor(s)/author(s) and do not necessarily represent those of editor(s) of the book.

All rights reserved. No part of this publication may be reproduced, stored or transmitted in any form or by any means, electronic, mechanical, photocopying, recording or otherwise, without the prior permission in writing of the publishers and the author.

All brand names and product names used in this book are trade names, service marks, trademarks or registered trademarks of their respective owners. The publisher and author are not associated with any product or vendor mentioned in this book.

Medical knowledge and practice change constantly. This book is designed to provide accurate, authoritative information about the subject matter in question. However, readers are advised to check the most current information available on procedures included and check information from the manufacturer of each product to be administered, to verify the recommended dose, formula, method and duration of administration, adverse effects and contraindications. It is the responsibility of the practitioner to take all appropriate safety precautions. Neither the publisher nor the author(s)/editor(s) assume any liability for any injury and/or damage to persons or property arising from or related to use of material in this book.

This book is sold on the understanding that the publisher is not engaged in providing professional medical services. If such advice or services are required, the services of a competent medical professional should be sought.

Every effort has been made where necessary to contact holders of copyright to obtain permission to reproduce copyright material. If any have been inadvertently overlooked, the publisher will be pleased to make the necessary arrangements at the first opportunity.

Inquiries for bulk sales may be solicited at: jaypee@jaypeebrothers.com

Practical Retinal OCT

First Edition: **2015**

ISBN 978-93-5152-532-5

Printed at Ajanta Offset & Packagings Ltd., New Delhi

Preface

Diagnoses must be the product of logical processes. Ocular imaging should be interpreted by logical methods of analysis and synthesis. I published analytical manuals on fluorangiography, indocyanine green angiography, OCT, Angio OCT and lately, "en face" OCT.

A rational method should be the basis of OCT interpretation. This manual illustrates a logical and simple analysis and interpretation method of OCT imaging, clearly stating the steps required to reach a diagnosis.

OCT sits next to the slit lamp in all ophthalmic offices and all ophthalmologists, optometrists, ophthalmic technicians and orthoptists must know how to use it. OCT highlights retinal and choroidal alterations in morphology, structure and reflectivity and facilitates the study of the various retinal layers, both separately and globally.

The images presented in this manual were recorded mainly with Optovue OCT (RTVue and XR Avanti OCT), and also with Heidelberg Spectralis. These instruments are both reliable and easy to use. There are high resolution optical sections, in sagittal view with the choroid visible to the sclera, with or without image averaging, or in frontal view, "en face", adapted the curvature of the fundus and other 3D images.

Illustrated with drawings, outlines, and sagittal and frontal ("en face") OCT images, this concise manual intends to show how to read and interpret OCTs, documenting and diagnosing the most common retinal pathologies. Simplified outlines are provided to facilitate the classification of morphological alterations.

Several tables offer guidance through the most difficult diagnoses.

Bruno Lumbroso
Marco Rispoli

Acknowledgments

I would like to thank Dr Marco Rispoli for his many years of invaluable collaboration, unerring imaging and image selection, and help with the chapters of this volume.

My thanks also go to Donata Piccioli, the graphic artist who has illustrated all of my scientific works and textbooks, for the precision of her beautiful drawings and the patience with which she has imaged my rudimentary schematic outlines.

My gratitude to Jay Wei, Bill Shields and Paul Kealey of Optovue (USA), who encouraged me to write the first edition of this practical and simple manual.

Bruno Lumbroso

Contents

Chapter 1 **OCT and Histology** 1
- Hyaloid Membrane and Vitreous *1*
- Inner Retina *1*
- Outer Retina *1*
- Photoreceptors, Cones and Rods *2*
- Retinal Support Structures *2*
- Retinal Pigment Epithelium *2*
- Bruch's Membrane *3*
- Choroid *3*

Chapter 2 **Normal Chorioretinal OCT Analysis** 11
- Retina and Choroid Morphology *11*
- Retina and Choroid Structure *11*
- Infrastructure *15*
- Vertical Structures *15*
- Horizontal Structures: Retinal Layers *16*
- Segmentation *16*
- Normal Tissue Reflectivity *18*
- High Reflectivity *18*
- Low Reflectivity *18*
- Choroid *18*
- Artifacts *19*

Chapter 3 **Qualitative Analysis of Pathologic OCTs** 21
- OCT Limits *22*
- OCT Imaging of Optic Disk *22*
- Morphology Study *22*
- Global Retinal Deformation *22*
- Retinal Profile and Retina Surface Deformation *23*
- Morphologic Posterior Alterations *24*
- Retinal Profile Outline *24*
- Outer Retina Outline *30*
- Retinal Structure Study: Segmentation *41*
- Structure: Segmentation *41*
- Inner Retina and Outer Retina *42*
- Outline–Segmentation *46*

- Reflectivity Study *48*
- High Reflectivity *48*
- Low Reflectivity *49*
- Abnormal Formations *55*
- Preretinal Formations *55*
- Intraretinal Formations *55*
- Posterior Formations *55*
- Shadow Areas (Screen Effect, Shadow Cone) *59*

Chapter 4 Quantitative Analysis of Pathologic OCTs 61
- Linear Measurements of Axial B-scans *61*
- Measurement of Coronal Images and "En Face" Scans *61*
- Quantitative Segmentation *62*
- Retinal Mapping *62*
- Quantitative Analysis: Thickness *63*
- Increased Thickness *63*
- Reduced Thickness *63*
- Thickness Variations of the Retinal Layers *65*
- Retinal Volume *65*
- Quantitative Choroid Analysis *66*
- Quantitative Optic Disk Analysis *66*

Chapter 5 "En Face" Imaging 67
- 3D Images: Segmentation of Chorioretinal Layers *67*
- Frontal Plane Scans *67*
- Frontal "En Face" Sections Fit to the Retinal Pigment Epithelium Concavity *68*
- Thickness of the Retinal Section *68*
- Tips and Parameters of the "En Face" Scanning Procedure *79*

Chapter 6 Synthetic Study 80

Chapter 7 Reporting 82
Reporting an OCT Test *82*
Scans *82*

Chapter 8 Glaucoma 85
- Glaucoma Scan Protocols *85*
- Glaucoma Follow-up *86*
- Ganglion Cell Complex Glaucoma Progression Report *86*
- Retinal Nerve Fiber Layer Glaucoma Progression Report *87*
- Normative Database *87*
- Glaucoma Diagnosis *87*

Index 93

Logical Method of OCT Analysis and Interpretation

The application of the classical Cartesian method of analysis to OCT is not self-evident. In fact, the images obtained with various imaging approaches are so complex and intricate that they cannot be simply considered a puzzle solved only by sorting.

OCT analysis must deconstruct the image into shapes, thickness and volume (morphology), and internal architecture, framework and stratification (structure). Furthermore, one must consider the interaction of high, medium and low reflectivity with both internal structure and morphology, as well as the effects of abnormal formations (fluid accumulations, exudates, hemorrhages, neoformations, etc.). These abnormal elements are bound by and adapt to the retinal and choroidal inner barriers, creating shadows that further modify scan appearances and cause reflectivity variations.

Retinal diseases can upset retinal architecture and morphology beyond recognition and interpretation. This in turn complicates any automatic segmentation attempts, whereas manual segmentation is error-prone and not always reliable. The quantification of morphological limits is as easy as they are distinct.

This manual states the basic principles of analysis. The drawings render inner architecture, framework and stratification. The outlines present a simplified version of common morphological features, both normal and pathologic, thus aiding their classification. High resolution optical sections, in sagittal view with the choroid visible to the sclera, with or without image averaging, or in frontal view, "en face", compensated for the curvature of the fundus, illustrate possible reflectivity and abnormal formations. OCT scans are in black and white, in order to acquire a higher number of details.

Numerous tables, organized by pathologic features (beige) and diagnostic indications (light blue), suggest at-a-glance the most frequent diagnoses, while reminding the less evident or rarer diseases.

Index of Tables

PATHOLOGIC FEATURES

Chapter 2
Table 1	Normal structure: reflectivity of normal tissues	19
Table 3	Retinal architecture	19

Chapter 3
Table 1	Qualitative analysis	22
Table 2	Morphologic alterations	38
Table 3	Retinal profile cross-section scan	39
Table 5	Nuclear and plexiform layers	43
Table 6	Inner nuclear layer	43
Table 7	Outer nuclear layer	43
Table 8	External limiting membrane	43
Table 9	External limiting membrane	43
Table 10	Inner/outer photoreceptor segment junction (ellipsoid zone)	44
Table 11	Inner/outer photoreceptor segment junction (ellipsoid zone)	44
Table 12	Retinal pigment epithelium	44
Table 13	Bruch's membrane	44
Table 18	Elevation or cystic space contents	50
Table 23	Low reflectivity structures	51
Table 26	Abnormal formations	56
Table 30	Abnormal structures and formations	58
Table 31	Analytic qualitative study of high reflectivity abnormal formations, by retinal level	59
Table 32	Shadow effect—screen effect	60

Chapter 4
Table 1	Quantitative analytical retinal study	66

DIAGNOSTIC INDICATIONS

Chapter 2
Table 2	Physiologic causes of choroidal reduced thickness	19

Chapter 3
Table 4	Causes of retinoschisis	42

Table 14	Lesions of outer layers and photoreceptors	45
Table 15	Subretinal deposits	45
Table 16	Differential diagnosis of isolated lesions of outer layers and photoreceptors	45
Table 17	Lesions of inner layers	45
Table 19	Common causes of macular edemas	50
Table 20	Causes of cystoid macular edemas	51
Table 21	Differential diagnosis of macular cystoid edema	51
Table 22	Diabetic maculopathies	51
Table 23	Causes of serous neuroretinal elevation	53
Table 24	Causes of serous retinal pigment epithelium detachment	54
Table 26	Causes of cotton wool exudates	56
Table 27	Causes of hard exudates	57
Table 28	Causes of neovascular membranes	58

Chapter 4

Table 2	Choroidal thickness variations	66

Chapter 5

Table 1	Clinical interest of 3D OCTs	68

CHAPTER
1
OCT and Histology

Optical coherence tomography (OCT) has allowed us to visualize in vivo the retinal structure. To understand OCT images it is indispensable to remember retina and choroid histology, even when a perfect parallelism is lacking.

In fact, some structures (photoreceptor inner/outer segment junction, ellipsoid, photoreceptor outer segment junction, retinal pigment epithelium cell villosities and interdigitations) are visible on OCTs because of their optical density, independent of their thickness, albeit they are not visible on histological sections.

HYALOID MEMBRANE AND VITREOUS

OCT shows vitreous humor, hyaloid membrane and the normal and pathologic structures (membranes and tractions), which can be clearly visible.

INNER RETINA

Inner plexiform layer, ganglion or multipolar cell layer and optic fiber layer form the ganglion cell complex or inner retina.

The internal limiting membrane is a thin membrane that adheres to nerve fibers and is formed by Müller's endfeet fiber ramifications.

Nerve fiber layer originating from the ganglion cells, the nerve fibers head for the optic nerve along an arched path. This layer is clearly hyperreflective as it is formed by horizontal structures.

The ganglion or multipolar cell layer consists of very bulky cells, the axons of which head for the eye's interior, forming the nerve fiber layer.

Inner plexiform layer connects bipolar and ganglion cells. It is slightly hyperreflective, due to its horizontal structures and, in terms of OCT segmentation, delimits inner and outer retina.

OUTER RETINA

Inner nuclear layer houses the horizontal bipolar cells nuclei, whose horizontal axons are located within the outer plexiform layer, and the Müller's cell nuclei. On OCT scans it is slightly hyporeflective.

Outer plexiform layer connects optic and bipolar cells. It includes the horizontal axons of the horizontal cells. On OCT scans it is slightly hyperreflective. It is formed by two sections: the inner one is dendritic, the outer one axonal, also known as Henle fibers.

PHOTORECEPTORS, CONES AND RODS

The photoreceptor layer includes the outer nuclear layer, consisting of visual cell nuclei. This layer forms a hyporeflective band, and is much thicker at the fovea level.

The photoreceptor cell body is slightly elongated. The nucleus completely fills the cell body. The protoplasm forms a conic foot on top, contacting the bipolar cells. The outer portion is divided in two: inner segment and outer segment. The latter is short and conical and comprises a stacked in succession series of disks. The inner portion is also divided in two: the inner myoid and the outer ellipsoid areas. The connection between the internal and external segments appears on OCTs as a hyperreflective horizontal band, not far from the retinal pigment epithelium-choriocapillaris complex. The new OCT nomenclatura gives it the appellation of ellipsoid.

The ellipsoid or junction line between the two segments is parallel to the retinal pigment epithelium band. Due to the increased length of the cones at the fovea level, this line is clearly distanced, at the foveola level, from the flat hyperreflective band constituted by the retinal pigment epithelium.

The external limiting membrane is very thin and forms a net surrounding the cone and rod base. Constituted by fibrillae originating from the Müller's fibers, the external limiting membrane is parallel to the ellipsoid.

RETINAL SUPPORT STRUCTURES

Müller's fibers are fibrous, vertical and very long supporting elements, which unite internal and external limiting membranes. Their nuclei are located within the bipolar cell layer, forming fans that become the external limiting membrane at one end and the internal limiting membrane at the other. Some horizontal branches of these fibers are part of the plexiform layer horizontal structure. Other important vertical structures include the cell chains consisting of photoreceptors linked to bipolar cells and subsequently to ganglion cells. Müller cells are Z shaped around the macula **(Figs 1 and 2)**.

RETINAL PIGMENT EPITHELIUM

The retinal pigment epithelium is formed by a layer of polygonal cells. The polygonal cell inner area is cup-shaped and it consists of villosities that receive the cone and rod tips. The nucleus is located in the outer cell area. Externally, the pigmented cell is in close contact with Bruch's membrane. On OCT scans, the retinal pigment epithelium seems to be constituted of 3 parallel bands: two relatively thick, hyperreflective bands, separated by a hyporeflective thin line. Several authors interpret the first hyperreflective layer as an interface between retinal pigment epithelium villosities and the photoreceptor external segment (interdigitation zone), while the second, more external layer is considered to comprise the retinal pigment epithelium cell bodies, together with their respective nuclei. Other authors think that the inner band corresponds to the tip of the photoreceptor external segment.

BRUCH'S MEMBRANE

Strong adhesions join retinal pigment epithelium, Bruch's membrane and choriocapillaris. Usually, the Bruch's membrane in not visible on OCT, but when there are drusen or small retinal pigment epithelium elevations, it appears as a fine horizontal line **(Figs 3 to 7)**.

CHOROID

The choriocapillaris is formed by polygonal vascular lobules, which receive blood from the short posterior ciliary arteries and drain it through venules into the vorticose veins. The main choroidal vessels are hyporeflective and can be seen in two layers: the Sattler's layer of small vessels and the Haller's layer of large vessels. Externally, at the sclera interface, the lamina fusca can be seen. A virtual space, i.e. the suprachoroidal space, separates the choroid from the sclera **(Figs 8 to 10)**.

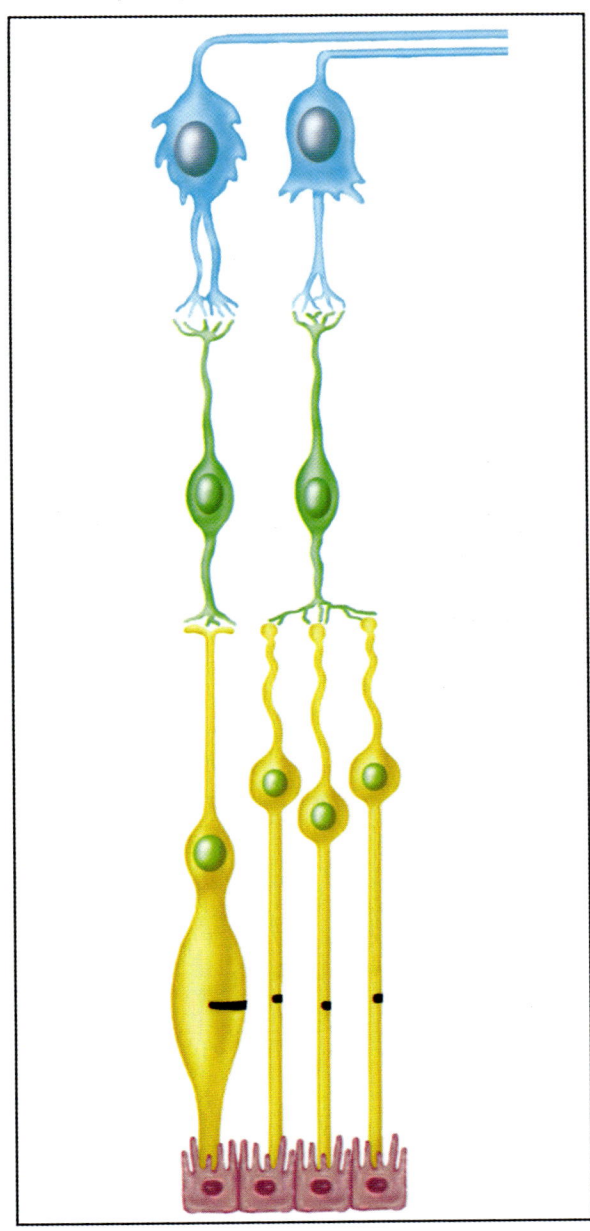

FIGURE 1: CELLULAR CHAINS
Retinal cells form chains of photoreceptors (cones or rods), linked to bipolar cells, which in turn, are connected to ganglion cells, whose axons constitute the optic nerve fibers.

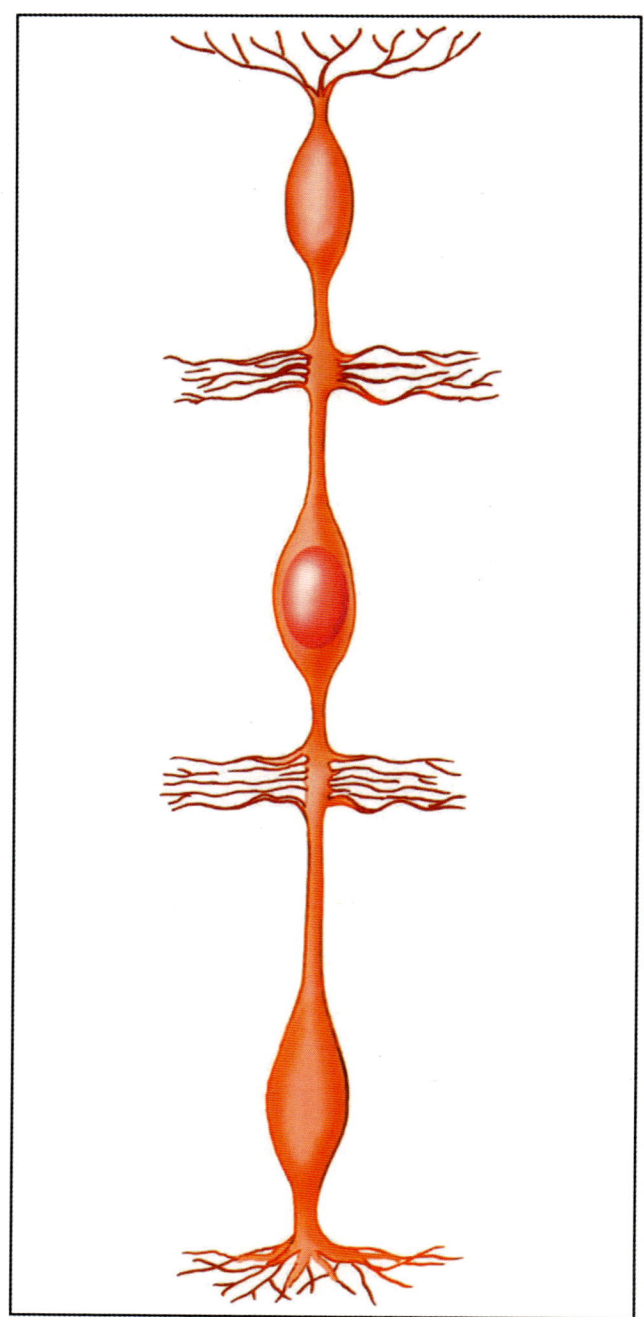

FIGURE 2: MÜLLER'S CELLS
The Müller's cells are vertical fibrous elements that join the internal and external limiting membranes. Their nucleus is located in the nuclear layer inside the bipolar cells. Fans of horizontal fibrillae project from the Müller's fibers, participating to the formation of the inner and outer plexiform layers.

OCT AND HISTOLOGY

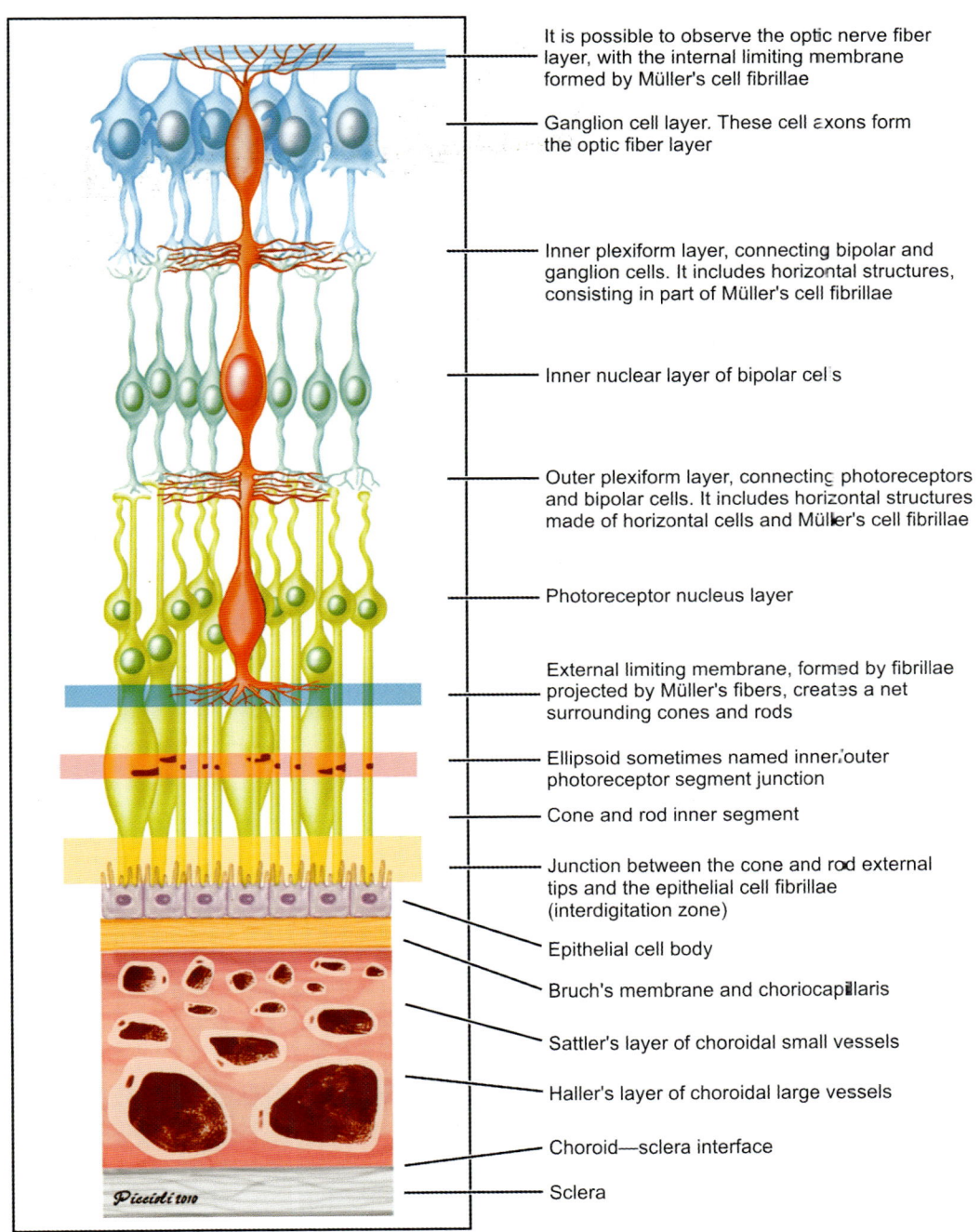

FIGURE 3: RETINAL AND CHOROIDAL LAYERS

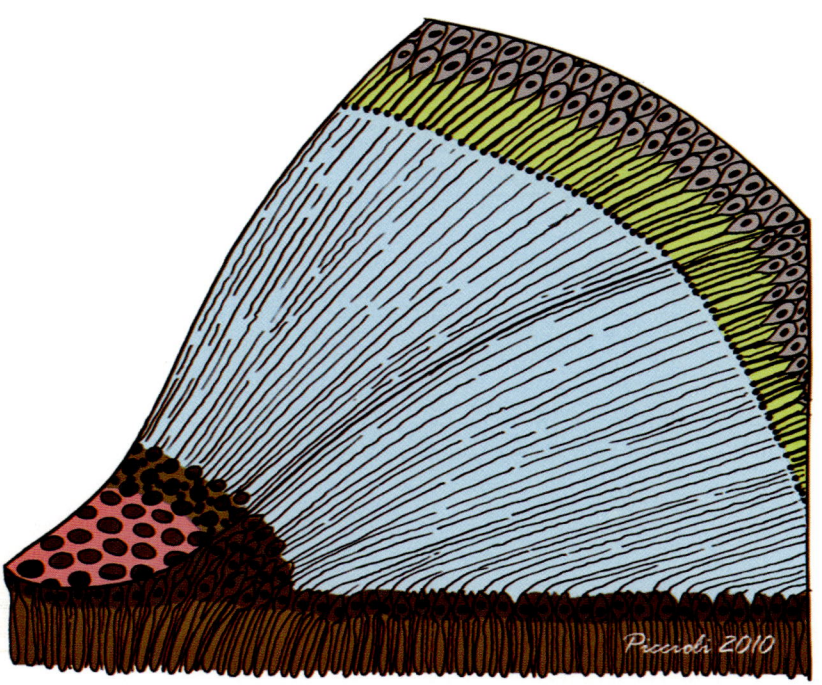

FIGURE 4: HENLE'S FIBER RADIAL STRUCTURE
At the fovea photoreceptor nuclei can be seen. From those nuclei originate the Henle's fibers, radially organized at 360 degrees, joining photoreceptors and bipolar cells. Thus a sunburst structure is formed, the infrastructure of which will determine the pseudocyst shape and layout within the cystoid edema ("flower" or "beehive" shaped).

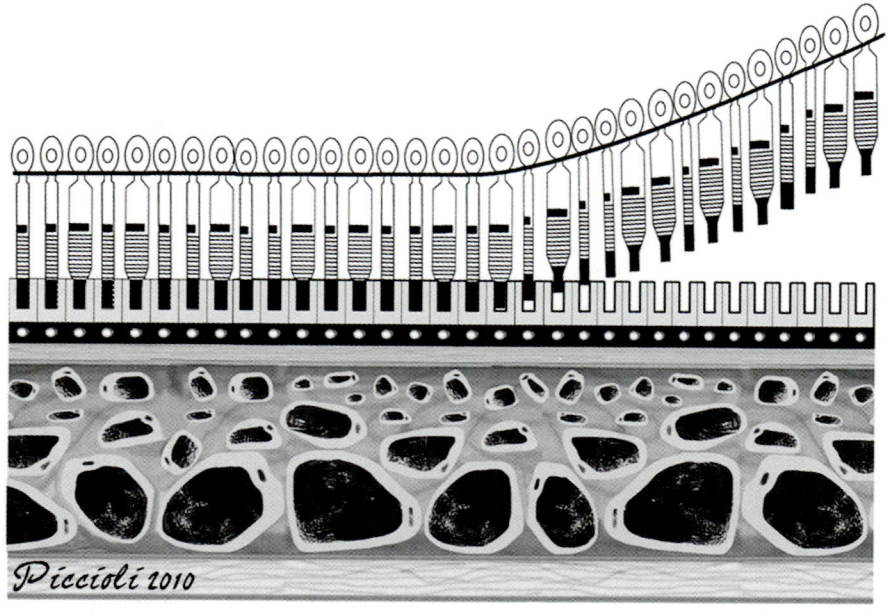

FIGURE 5: SEROUS DETACHMENT OF NEUROEPITHELIUM
A serous detachment of neuroepithelium in a central serous chorioretinopathy (CSCR) case highlights the level in which the detachment of photoreceptors from the retinal pigment epithelium takes place. According to one of the theories generally accepted, the separation between neuroretina and retinal pigment epithelium happens at the junction between the photoreceptor external tips and the retinal pigment epithelium villosities (interdigitation zone).

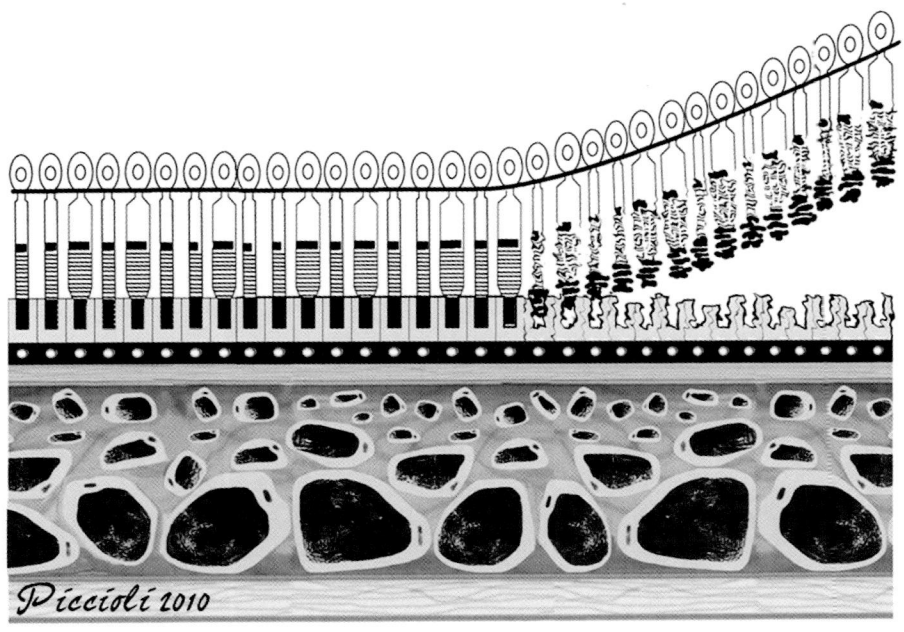

FIGURE 6: SEROUS DETACHMENT OF NEUROEPITHELIUM WITH ELEVATED PHOTORECEPTOR ALTERATIONS

Long-term serous detachment of neuroepithelium in case of chronic CSCR. Note the alterations shown by the photoreceptors, elevated for over 6 months. Photoreceptor discs are edematous, turgid and altered. Retinal pigment epithelium villosities also show clear alterations.

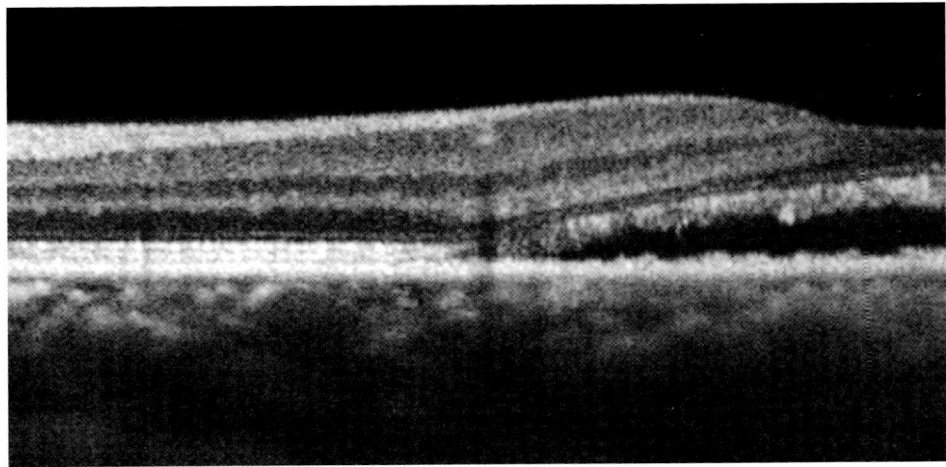

FIGURE 7: SEROUS DETACHMENT OF NEUROEPITHELIUM WITH ELEVATED PHOTORECEPTOR ALTERATIONS

The retinal pigment epithelium—choriocapillaris complex is often subdivided into three parallel bands: two are relatively thick and hyperreflective, separated by a narrow hyporeflective line. In this case of acute serous chorioretinopathy, the internal hyperreflective band constitutes the interface between the photoreceptor outer segment and the retinal pigment epithelium cells. The photoreceptor outer segment tip is separated from the external band that includes retinal pigment epithelium cell nuclei, pigment granules, Bruch's membrane and the choriocapillaris. Above the serous elevation, we can see the elevated internal band; and further up the ellipsoid (photoreceptor inner/outer segment junction) and the external limiting membrane. Immersed within the serous elevation fluid, the photoreceptor outer segment appears altered, due to of cone and rod discs engorgement and alterations. Irregularities can instead be seen at retinal pigment epithelium level.

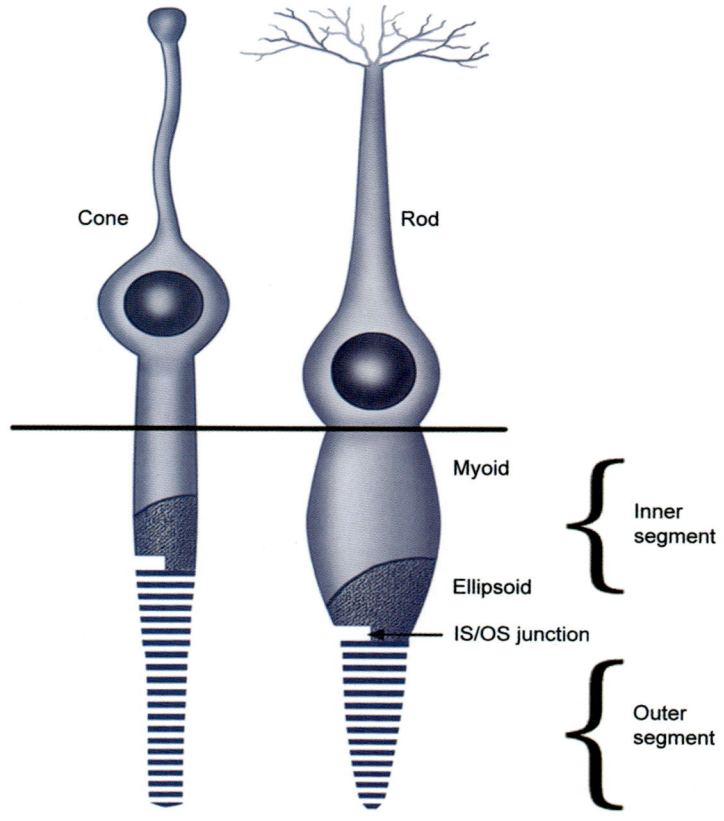

FIGURE 8: CONE AND ROD
Between the external limiting membrane and the ellipsoid zone there is the myoid zone of the inner segment of the photoreceptors.
   The junction between the inner and outer segments of the photoreceptors is now called ellipsoid as it has been demonstrated that it corresponds to the portion of the inner ellipsoid segment of the photoreceptors that is markedly hyperreflective.

OCT AND HISTOLOGY

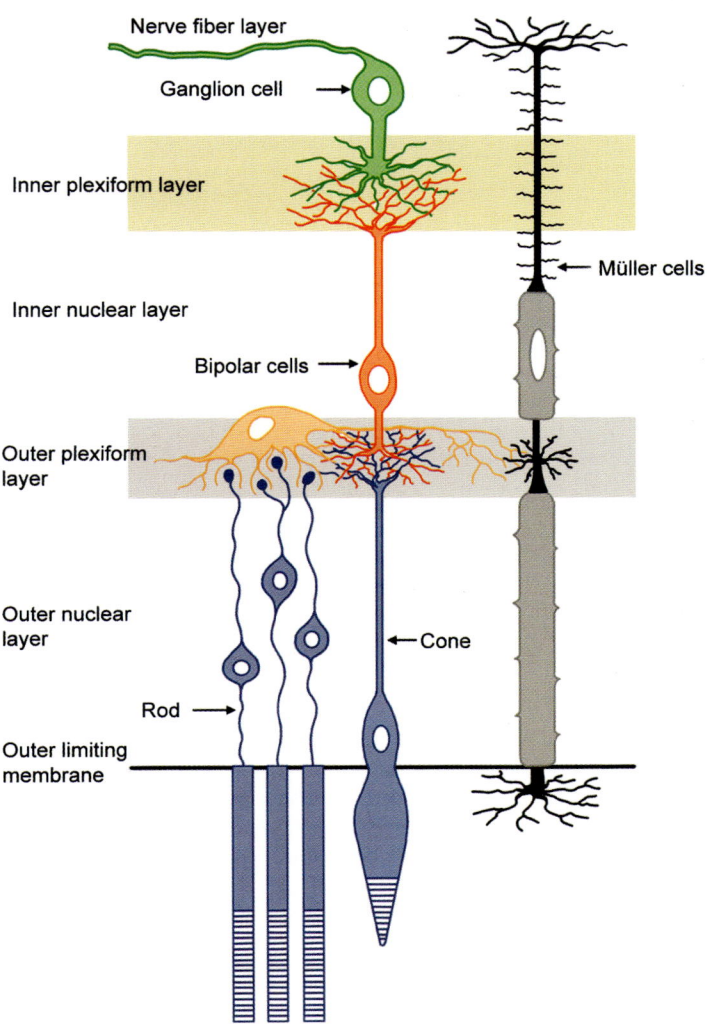

FIGURE 9: RETINAL LAYERS
Müller cells are the most important retinal support and contribute in the formation of the inner and outer limiting membranes, and of the inner and outer plexiform layers.

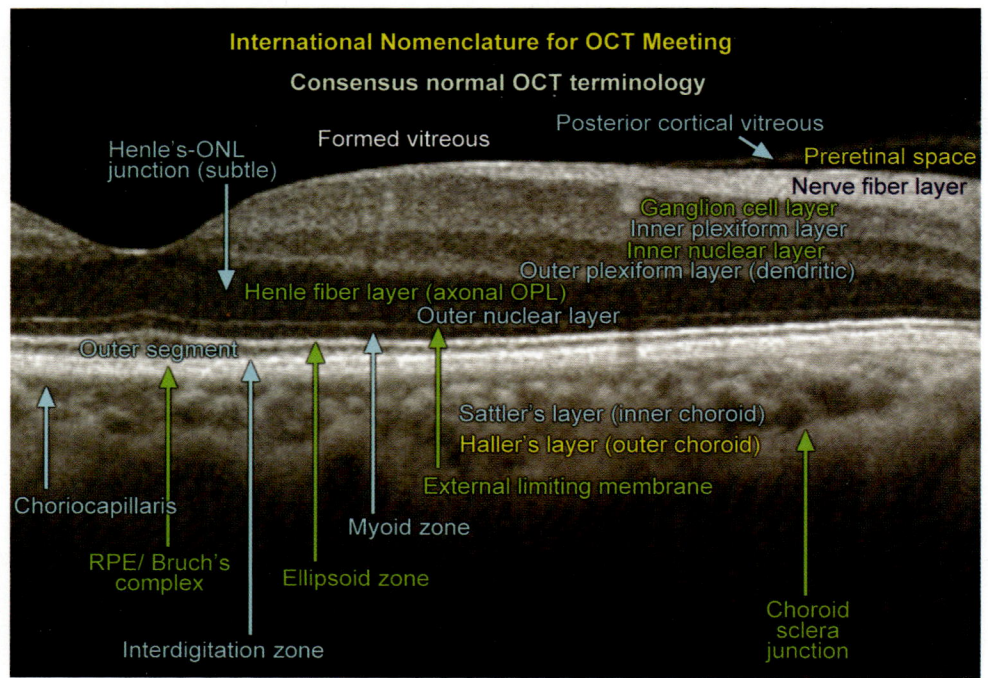

FIGURE 10: NEW OCT TERMINOLOGY. A NEW OCT CLASSIFICATION OF RETINAL AND CHOROIDAL LAYERS IS NOW INTERNATIONALLY USED. IT IS THEREFORE DESIRABLE THAT THE NEW TERMINOLOGY BE USED FOR OCT ANALYSIS, INTERPRETATION AND REPORTS

The terminology regarding the inner layers has remained unchanged down to the inner nuclear layer.

The outer plexiform layer is now called outer plexiform layer, *dendritic part*.

The outer nuclear layer is divided into two parts. The first is less reflective and is called outer plexiform layer, *axonal part (Henle fibers)*. The outer nuclear layer second part, that is slightly more reflective, is now called outer nuclear layer. The boundary between these two layers is very subtle and often difficult to see but becomes sharper when the incidence of the ray is slightly more oblique. This line forms the boundary between Henle's fibers and the outer nuclei.

The external limiting membrane remains unchanged.

Between the external limiting membrane and the ellipsoid zone there is the myoid zone of the inner segment of the photoreceptors.

The junction between the inner and outer segments of the photoreceptors is now called ellipsoid as it has been demonstrated that it corresponds to the portion of the inner ellipsoid segment of the photoreceptors that is markedly hyperreflective.

Between the ellipsoid and the interdigitation zone lies the outer segment of the photoreceptors.

The first hyperreflective layer of the pigment epithelium is called interdigitation zone because it consists of the interdigitation between the cells of the pigment epithelium and the external extremities of the photoreceptors.

A hyperreflective strip is located more externally, that corresponds to the pigment epithelium and to Bruch's membrane.

Underneath it, in the following order we see the choriocapillaris, Sattler's layer, Haller's layer and the choroid-sclera junction.

(*Courtesy* Giovanni Staurenghi MD)

CHAPTER
2
Normal Chorioretinal OCT Analysis

The normal optical coherence tomography (OCT) analysis requires three basic steps:
- Morphology study
- Retinal and choroidal structure study
- Reflectivity study.

In order to study them, OCT images are preferably viewed in grayscale rather than in conventional colors. This allows highlighting minimal density variations and seeing details that could not be seen otherwise. Negative images can help to better visualize some structures. The colors of OCT color images are arbitrarily chosen by the software, which associates a given color to each degree of reflectivity: thus marked color differences are observed, whereas in reality there is a graduality of reflectivity **(Figs 1A to C)**.

Retina and Choroid Morphology

The morphology study includes assessment of shape and profile of vitreous-retina and retina-choroid interfaces, as well as the choroid-sclera interface. A morphology study thus requires the evaluation of the shape and dimensions of both retina and choroid.

Retina and choroid have a concave shape within the sclera, similar to a bowl. The vitreous-retina interface does not have a regular curvature (paraboloid shape). The fovea forms a depression surrounded by an area that is thickened by the nuclei of ganglion cells and inner nuclear layer cells.

The vitreous adheres clearly to the optic disc margin and (in juvenile subjects) to the fovea.

The adhesion decreases with age.

Retina and Choroid Structure (Figs 2 and 3)

Retina and choroid are formed by multistratified tissues with parallel layers. The retinal structure includes, in addition to the parallel layers, transversal formations that join the various layers together.

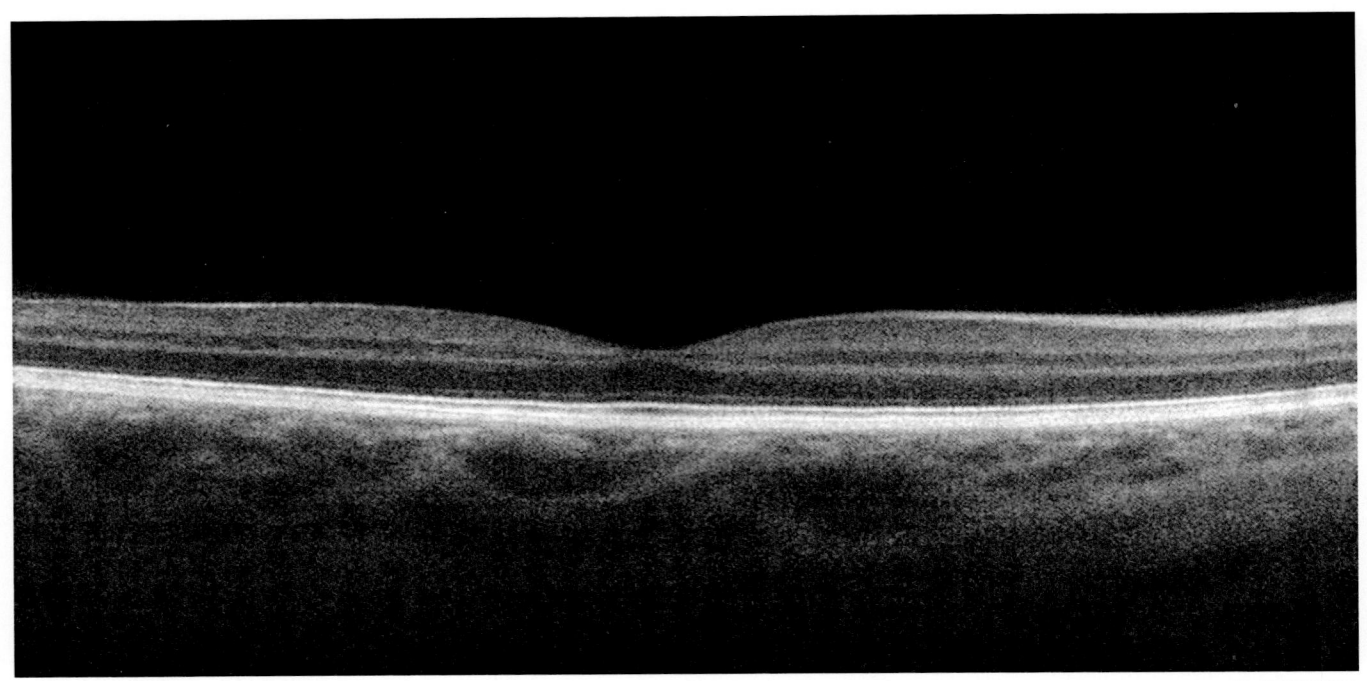

FIGURE 1A

FIGURE 1B

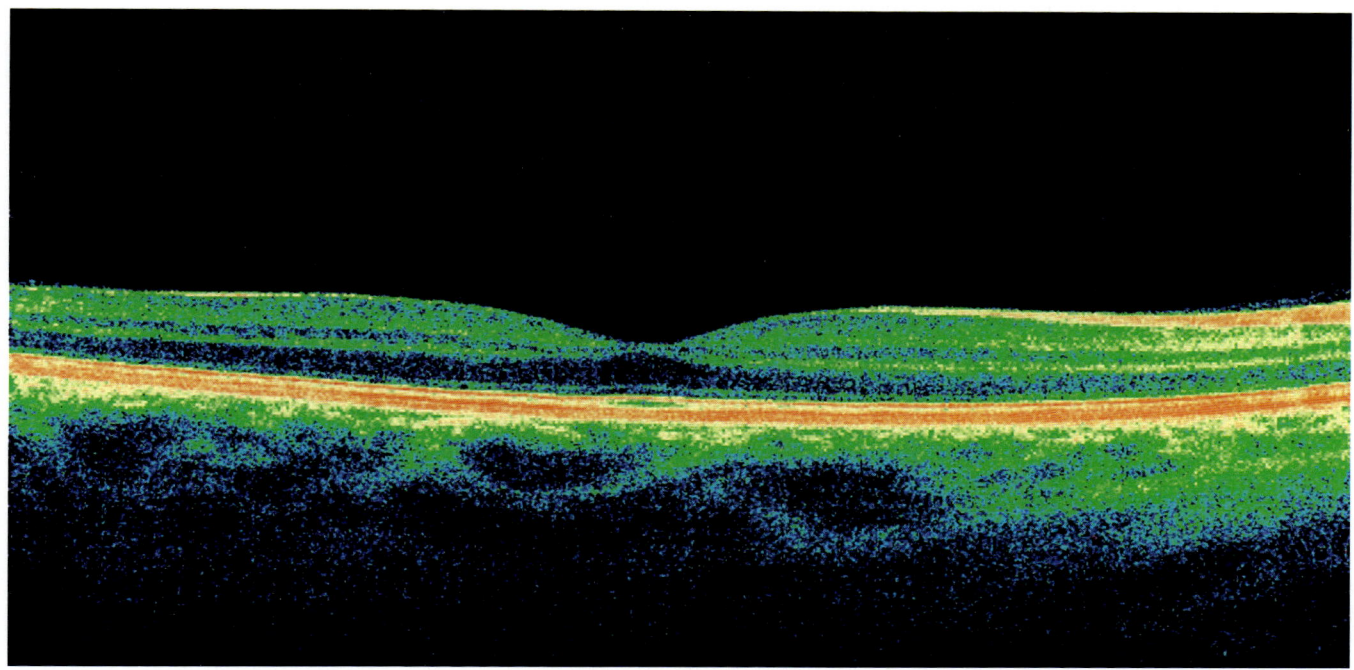

FIGURE 1C

FIGURES 1A TO C: OCT OUTLINE OF RETINAL STRUCTURE IN GRAYSCALE, NEGATIVE AND CONVENTIONAL COLORS

Grayscale and black and white allow viewing the image in greater detail, while colors offer greater contrast. In general, grayscale and negative black and white are preferred to study the more difficult cases.

The retinal structures are visible on histologic sections, except the inner/outer photoreceptor segment junction and the interface between the outer photoreceptor segment and the villi of retinal pigment epithelium cells, which are derived from the tissue optical and electronic characteristics. It is now accepted that the inner/outer photoreceptor segment junction (ellipsoid) corresponds to the ellipsoid part of the photoreceptors.

Located within the ellipsoid (junction between inner/outer photoreceptor segments) and the photoreceptor bodies, the external limiting membrane can be seen on histologic sections.

Three structures: the ellipsoid (inner/outer photoreceptor segment junction); the interface between outer photoreceptor segment and retinal pigment epithelium; and the external limiting membrane are visible on OCT Spectral due to their optical and electronic properties.

Also note the shadow projected upon the posterior layers by normal retinal vessels.

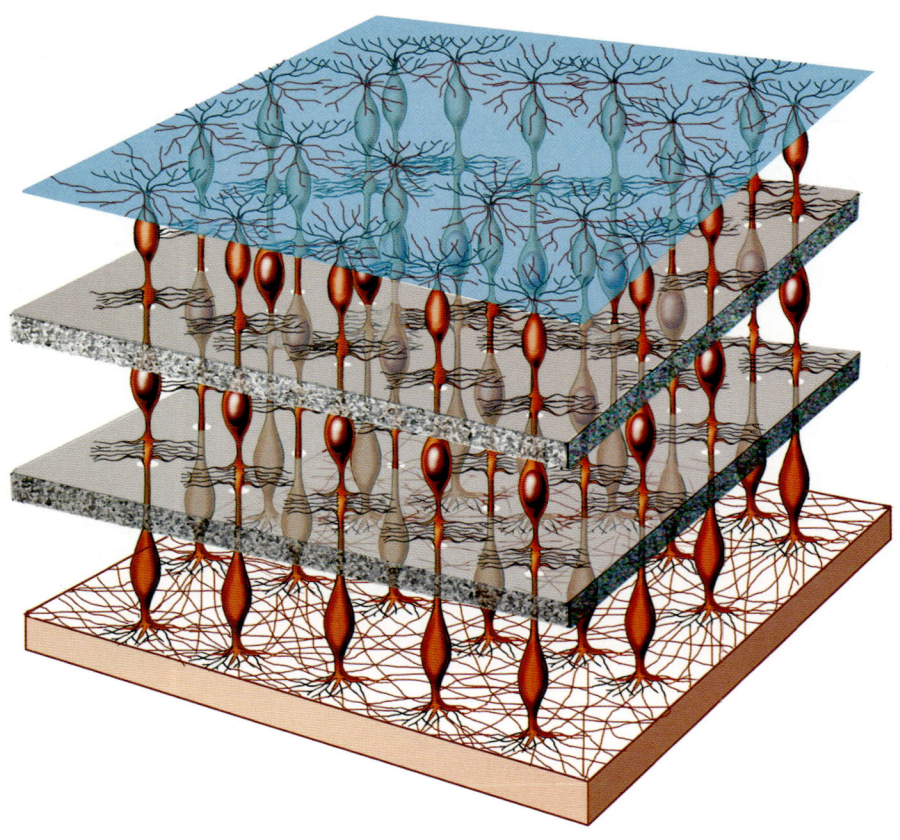

FIGURE 2: RETINA SUPPORT STRUCTURE

Horizontal structures: Inner and outer plexiform layers, external and internal limiting membranes.

Müller fibers fan out top and bottom and becoming the external limiting membrane at one end and the internal limiting membrane at the other. The inner and outer plexiform layers consist of horizontal and amacrine cells and an intercellular synaptic connection network between bipolar cells and photoreceptors on one side and bipolar and ganglion cells on the other. The outer plexiform layer is also formed by an axonal part, the Henle fibers. The plexiform layers do not constitute true histologic membranes, as much as effective barriers, albeit less strong, resistant and impermeable than the internal and external limiting membranes. A complex fiber network forms horizontal barriers against the spreading of fluid through the retina. The inner plexiform layer is more resistant and less permeable than the outer one.

Vertical structures: Müller fibers and cell chains consisting of photoreceptors linked to bipolar and ganglion cells.

Müller fibers are fibrous, vertical and very long supporting elements, which unite internal and external limiting membranes. Their nuclei are located within the bipolar cell layers. Other important vertical structures include the cell chains consisting of photoreceptors linked to bipolar cells and subsequently to ganglion cells.

These vertical and horizontal structures limit the expansion of pathologic lesions, exudates, hemorrhages and edema. Thus, the retinal tissue forms elongated cavities that appear on OCTs as pseudocystic holes.

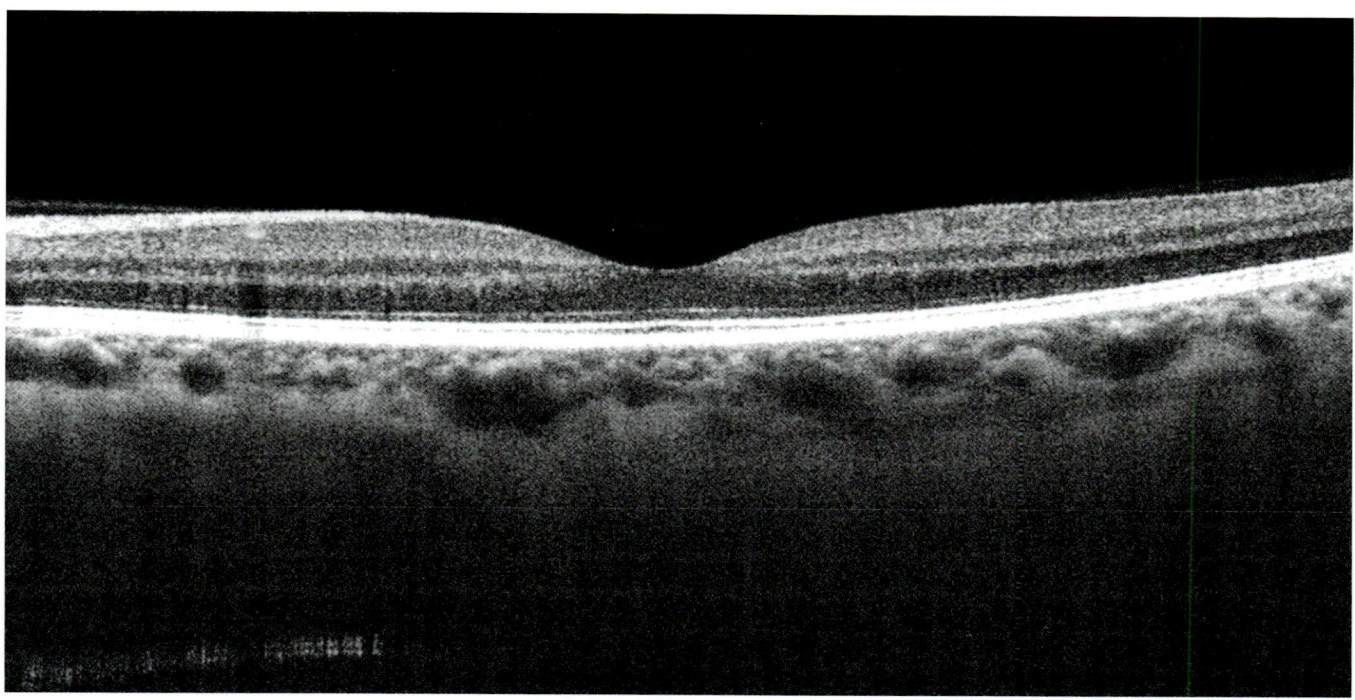

FIGURE 3: NORMAL
Retina and choroid are multistratified membranes. The B-scan images of the normal retina show nerve fibers, ganglion cells, inner plexiform layer, inner nuclear layer of bipolar and horizontal cells, and outer plexiform layer.
 Next, the outer nuclear layer, which contains the photoreceptor nuclei, external limiting membrane, inner/outer photoreceptor segment junction, retinal pigment epithelium cells, Bruch's membrane, and choriocapillaris, choroid divided in Sattler's layer of small vessels and Haller's layer of large choroidal vessels, and the sclera.
 An important element is the ellipsoid that was named junction or articulation between the inner and outer photoreceptor segment. Also evident is the external limiting membrane, parallel to the inner/outer photoreceptor segment junction, and placed between this junction and the photoreceptor cell bodies.

INFRASTRUCTURE

The retina is supported by a complex scaffold, made of capillary fibers and organized cells, which constitute true barriers to fluid diffusion. Retinal cohesion is maintained by horizontal and vertical formations. These vertical (cell chains and Müller cells) and horizontal (plexiform layers, limiting membranes) formations explain the location, dimensions and shape of exudates, hemorrhages and cystoid cavities visualized by OCT. Anatomic barriers vertically and horizontally block the spreading of pathologic processes.

VERTICAL STRUCTURES

Müller's cells join the internal limiting membrane with the external one, extending through two plexiform layers. Other vertical structures include the cell chains formed by photoreceptor cells linked to bipolar cells, linked in turn to ganglion cells.

HORIZONTAL STRUCTURES: RETINAL LAYERS

The internal and external limiting membranes are easily recognizable on retinal histologic sections. The internal limiting membrane is formed by Müller's cell fibrillae.

The inner and outer plexiform layers consist of horizontal and amacrine cells and an intercellular synaptic connection network between bipolar cells and photoreceptors on one side and bipolar and ganglion cells on the other. They do not constitute true histologic membranes, as much as effective barriers, albeit less strong, resistant and impermeable than the internal and external limiting membranes. The plexiform layers consist of a complex and tangled network of fibers, forming effective horizontal obstacles to the diffusion of fluid through the retina. The inner plexiform layer is more resistant and less permeable than the outer one. Part of the outer plexiform is formed by horizontal Henle fibers. Henle's fibers join the cones nuclei with the bipolar cells.

These vertical and horizontal structures limit the spread of pathologic lesions, exudates, hemorrhages and edema and contribute to the determination of the pseudocystic cavities shape.

At the fovea level, Henle's fibers form a sunburst structure in the foveal area, which is clearly shown in a retinal frontal section and in flat mounted histologic sections. The cones are placed at the center and around them it is possible to see the photoreceptor nuclei. Henle's fibers join the cones nuclei with the bipolar cell nuclei at the fovea periphery. In the foveal area, Müller's cells are Z shaped and diagonally join the internal to the external limiting membranes. Henle's fibers contribute to determine the flower or beehive shape of the pseudocystic cavities typical of macular cystoid edema.

Outside the foveal area, the nerve fibers are arcuate **(Fig. 4)**.

SEGMENTATION (FIGS 5 AND 6)

Retina and choroid are formed by stratified structures, constituted in turn by stacked layers having different reflectivities. Therefore, segmentation techniques can show the stratified structure of the eye membranes, analyzing the bands of homogeneous reflectivity, both high and low. Image segmentation allows singling out and highlighting a group of segments that, once put together, reconstruct the image itself. This regularly stratified structure is modified and sometimes subverted by retinal pathology.

The retina can be divided into the inner and outer retina. The inner retina includes a layer of nerve fibers and a layer of ganglion cells. The inner plexiform layer serves as a border between the inner and outer retina.

The outer retina includes inner nuclear layer, outer plexiform layer, external limiting membrane, inner/outer photoreceptor junction segment and retinal pigment epithelium complex.

Outer plexiform layer is divided in a dendritic layer and an axonal (Henle fibers) layer.

Many current OCTs allow the subdivision (segmentation) of the retina into its different layers, structures and most important components. The nerve fiber layer segmentation has been the first segmentation automatically performed by all OCT instruments. The ganglion cell layer segmentation is also essential to studying glaucoma and neurodegenerative diseases and their evolution.

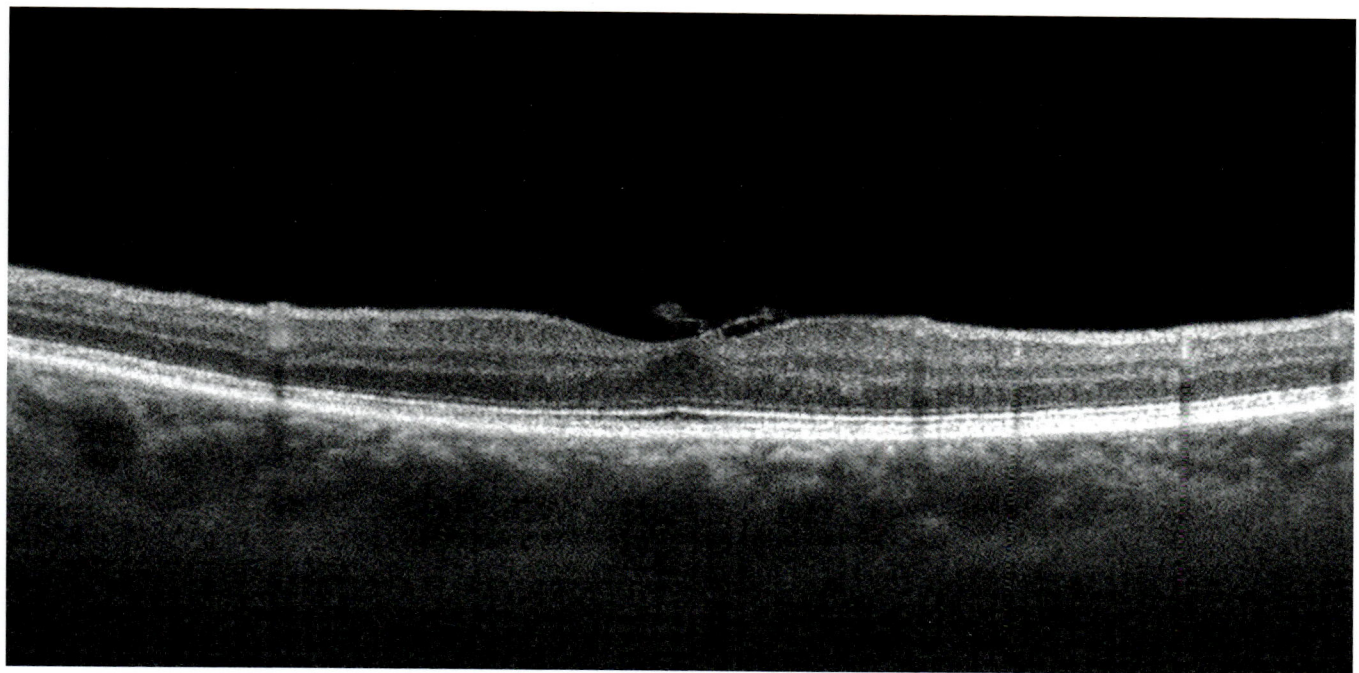

FIGURE 4: RETINAL PIGMENT EPITHELIUM/CHORIOCAPILLARIS COMPLEX
Retinal pigment epithelium/choriocapillaris complex can be divided in three parts: two large hyperreflective bands, separated by a thin hyporeflective line.
 The inner hyperreflective area corresponds to photoreceptor external tips or to the articulation between photoreceptors tips and retinal pigment epithelium villosities (interdigitation zone). The external hyperreflective band corresponds to retinal pigment epithelium cell nuclei, pigment granules, Bruch's membrane and choriocapillaris.

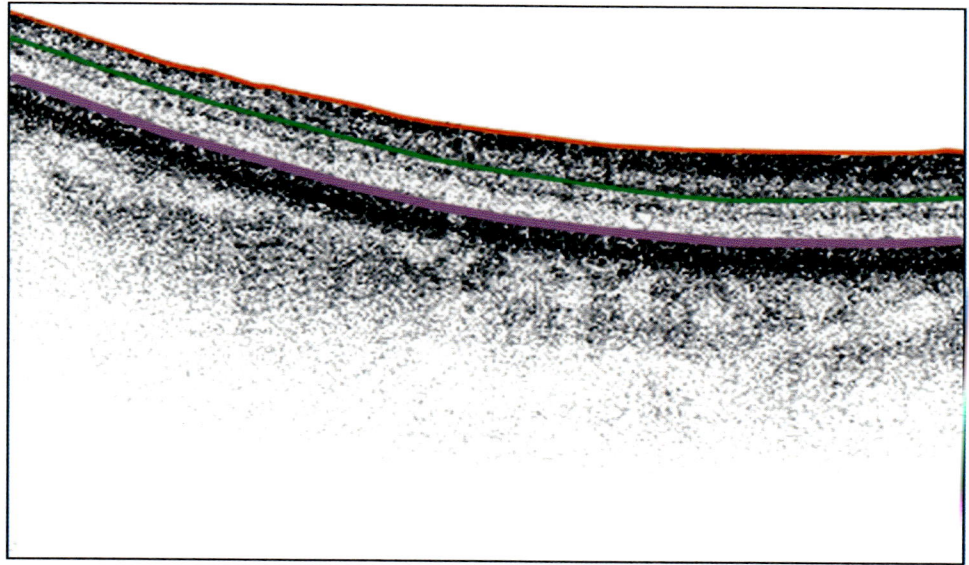

FIGURE 5: SEGMENTATION
The green line located at the level of the inner plexiform layer marks the division or segmentation between outer and inner retina.

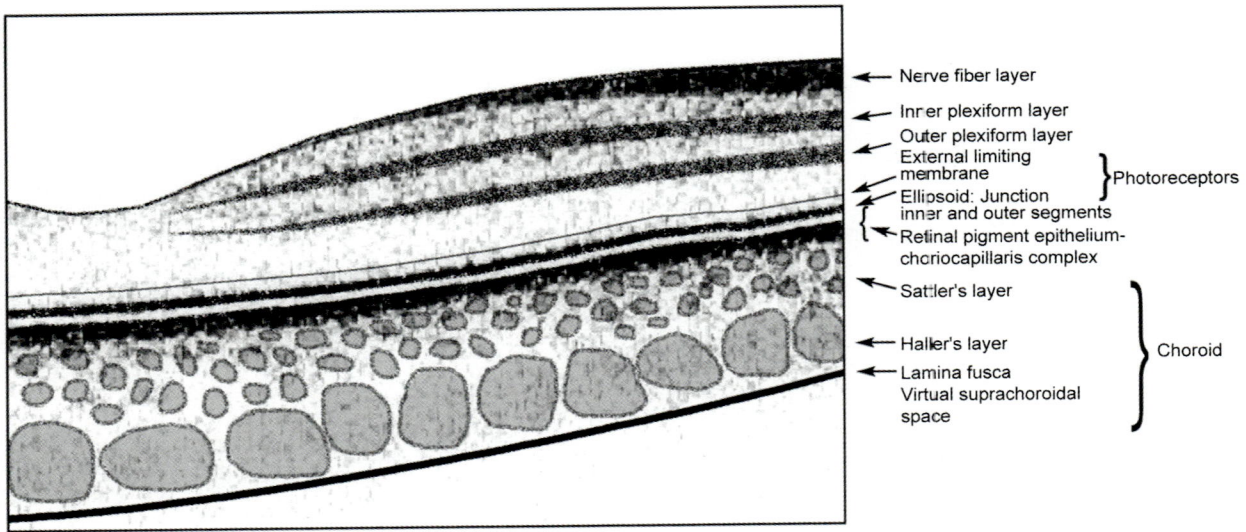

FIGURE 6: SEGMENTATION
The figure shows all retinal horizontal layers, retinal pigment epithelium, choroid and sclera. Although all current OCT instruments cannot segmentate all of retina's layers, there are some commercial devices already capable to do so. In all likelihood, all commercial instruments will make segmentation available to the clinicians.

Normal Tissue Reflectivity

HIGH REFLECTIVITY

The strength of the light signal as reflected by a tissue depends on its optical density and the light absorption capability of the tissues that are located in front of it.

Reflectivity depends on:
- The quantity of light that reaches a given layer, after being absorbed by the tissues it crosses.
- The quantity of light that a layer can reflect.
- The quantity of reflected light that reaches the detector, after being further absorbed by the tissues it crosses.

Vertical structures such as the photoreceptors are less reflecting than horizontal ones such as nerve fibers and plexiform layers.

LOW REFLECTIVITY

Low reflectivities can be caused by a reduced reflectivity of an atrophic tissue, the presence of vertical structures (photoreceptors) and fluid-containing areas, especially in pathologic OCTs **(Table 1)**.

CHOROID

Choroidal vessels appear hyporeflective and can be divided in two layers: the Sattler's layer of small vessels and the Haller's layer of large vessels. The intravasal connective tissue reflectivity is mostly medium, but can occasionally be high. Externally, the lamina fusca appear as a thin line. Normally, the suprachoroidal space is not visible.

TABLE 1 — Normal structure: Reflectivity of normal tissues

High Reflectivity
- Nerve fiber layer
- Inner/outer photoreceptor segment junction
- External limiting membrane
- Retinal pigment epithelium/choriocapillaris.

Medium Reflectivity
- Plexiform layers.

Low Reflectivity
- Nuclear layers
- Photoreceptors.

TABLE 2 — Physiologic causes of choroidal reduced thickness

- Age
- Myopia.

TABLE 3 — Retinal architecture

Horizontal barriers
- Internal limiting membrane
- Inner plexiform layer
- Outer plexiform layer
- External limiting membrane
- Retinal pigment epithelium
- Bruch's membrane
- Suprachoroidal space
- Sclera.

Vertical formations
- Müller's cells
- Vertical cell chains formed by photoreceptors, bipolar and ganglion cells.

Usually, the choroid has a thickness of about 300 microns, which decreases with age starting at age 30 years. Choroid thickness decreases with the increase of age and with myopia **(Tables 2 and 3)**.

Artifacts

Modern OCT devices acquisition strategy makes artifacts most unlikely. In B-scan, the system averages a variable number of images. Therefore, a possible movement or blink of the patient's eye can cause only a few abnormal scans, which will be excluded from averaging. Setting the image overlap, the tolerance value can be defined in inverse proportion to the desired image clarity.

When scans are performed to calculate the macular cube, the software automatically compensates the alignment of serialized scans, within a given limit, beyond which the images are offset. The patient needs to remain still for about 4 seconds in order to properly acquire the macular cube. In newest devices acquisition is faster. Relevant artifacts can be observed at the level of vertical control scan. This scan is the product of a reconstruction based on A-scans within horizontal scans. For this reason, micromovements along the anteroposterior axis can generate an abnormal vertical B-scan. To solve this problem, immediately after the cube acquisition, the system presents a coronal view of the retinal pigment epithelium layer. If that view present aberrations, reacquire the image.

The last aberration typology is represented by the hyporeflectivity of some structures, usually caused by the opacity of dioptric media or, more rarely, by an incorrect alignment. In these cases, the marker positioning (external limiting membrane and retinal pigment epithelium) could be incorrect. Expert users can manually reposition the markers, obtaining an acceptable topography.

CHAPTER 3

Qualitative Analysis of Pathologic OCTs

OCT reading should be performed in two steps:
- Qualitative and quantitative analytical study
- Synthetic study.

In the first analytical phase we subdivide and analyze OCT various elements. To this effect we need to perform segmentation, sagittal (B-scan) and frontal ("en face") scans, studying separately morphologic alterations, structural alterations, hyperreflectivity, hyporeflectivity and quantitative alterations.

Only after these elements are studied and analyzed we can move to second stage, the synthesis, using the results of the physical examination, anamnesis, fluorangiography, indocyanine green fluorangiography, autofluorescence imaging. The confrontation and synthesis of all these elements will allow interpreting the OCT and reaching a diagnosis.

The analytical step, first and important step, can be subdivided in **(Table 1)**:
- **Qualitative analysis**
  - Morphology study
  - Structure study: segmentation
  - (Hyper-, hypo-) reflectivity study
  - Study of abnormal formations and deposits
  - Study of shadow areas
- **Quantitative analysis.**

OCT is easy to use and the learning curve is short; it is a reliable, sensitive test (its resolution equals 3 microns). The images thus generated can be analyzed, quantified, saved and compared to those obtained by subsequent tests, allowing one to diagnose, quantify the lesions and follow up the illness progress.

The images are reproducible with different instruments and by different operators; the scanning is fast, simple and most importantly does not require the administration of contrast media. It is a noninvasive examination to be preferred to other more invasive examination tests that can induce complications and cause medical-legal problems.

OCT LIMITS

A good transparency of the dioptric media and a normal tear film or artificial tears are needed. The test is difficult to perform in the presence of high myopia, nubecolae or corneal edema, lens opacity, vitreous turbidity, vitreal hemorrhages.

Currently, scans can generally be performed at the posterior pole level, although today's thanks to fast technical progress there are OCT devices capable to explore the very large retina segments (XR Avanti, Optovue).

TABLE 1	Qualitative analysis
The qualitative analysis of pathologic OCTs includes five basic steps: 1. **Morphology study**, i.e. the study of shape and profile of retina-retina and retina-choroid interfaces, as well as the choroid-sclera interface. 2. **Structure study** (segmentation). Retina and choroid are multistratified tissues, with layers that are parallel to each other and a concave bowl-like shape within the sclera. The retinal structure also includes the transversal formations that join together the various layers. 3. **Reflectivity study** (high, low reflectivity). 4. **Study of abnormal formations and deposits**. 5. **Study of shadow areas**.	

OCT IMAGING OF OPTIC DISK

The horizontal scans shows the physiological excavation in which the high reflectivity nerve fibers penetrate. The retina stops at the border with the papilla. The retinal pigment epithelium choriocapillaris is abruptly truncated, while superficial nerve fibers merge with those of optical nerve. The scleral ring can be observed.

A quantitative analysis allows one to establish the optic disk diameter and the diameter and area of excavation at various depth levels.

The subject will be dealt with in detail in the chapter "Papillary Pathology and Glaucoma".

Morphology Study

Morphology: The determination of the shape and dimension of retina and choroid or the parts thereof. Proceeding from the vitreous toward the choroid, a morphology study requires the layer evaluation of retina and choroid or any parts thereof.

GLOBAL RETINAL DEFORMATION

Concavity (myopia): OCTs highlight a marked concavity in case of strong myopia and, especially, posterior staphyloma.

Convexity: It is often seen in case of dome-like detachment of the retinal pigment epithelium, but it can be caused by a subretinal cyst or tumor. In the latter case, the convexity is more flat and encompasses the subretinal layers (epithelium and choriocapillaris), lacking the slant observed in epithelium detachment. In most cases, it is impossible to locate the tumor itself. Almost always, the neuroepithelium presents localized edema or elevations. Another possible cause is dome-shaped macula, frequently observed in myopia and generally due to scleral thickening.

A retinal localized concavity suggests scleral staphyloma in the myopic eye or scleritis outcome.

RETINAL PROFILE AND RETINA SURFACE DEFORMATION (TABLES 2 AND 3)

The disappearance of the foveal depression frequently signals a clinically significant retinal edema.

Epiretinal membranes exert vertical or horizontal traction onto the retinal surface, deforming its profile, sometimes creating intraretinal cracks. They always induce retinal folds formation.

Macular pucker produces visible surface irregularities. The retinal surface is deformed by folds and curled under the pull exerted by the epiretinal membranes. Said membranes, often visible, can be coalescent, unidentifiable from and confused with the nerve fiber layer. They induce retinal folds formation. "En face" scans visualize well these retinal fods, which often are star-shaped or parallel waves **(Figs 1 to 6)**.

Macular pseudohole: An augmented foveal depression, caused by epiretinal membranes, simulates a macular hole in the retina. The residual retinal tissue on the depression bottom is deformed, but not decreased.

Lamellar hole: The foveal depression is increased by the disappearance of internal retinal layers. The partial presence of the retinal tissue above the retinal pigment epithelium suggests the diagnosis of lamellar hole **(Fig. 7)**.

Macular holes: The capability to identify the macular hole, classify it and measure its diameter constitutes an important progress achieved by OCTs.

According to Gass' classification, we can observe:
- Stage 1: Disappearance of the foveal depression and formation of a small area, optically empty (cyst) below the retinal surface.
- Stage 2: The inner retina is partially fissured, the operculum still adheres to the hole's edge and there is a slight increase in thickness.
- Stage 3: The hole has reached full retinal thickness, all the way to the retinal pigment epithelium, the operculum has lost adherence to the retina, a retinal edema is present, the retinal thickness has increased and presents a small detachment.
- Stage 4: The hole interests full retinal thickness, there is a loss of retinal tissue, a marginal retinal cystoid edema, and the margins are elevated.

These are the typical features seen when the B-scan goes through the hole center. When the scan is not perfectly centered and crosses the lesion margins, appearance can be deceptive, suggesting a pseudohole or a hole in progress.

MORPHOLOGIC POSTERIOR ALTERATIONS

The retinal pigment epithelium can present a decreased or increased thickness. Sometimes, that thickness can be irregular, due to the alteration or loss of one or two layers. Layers can appear abnormally and irregularly thickened or disaggregated.

The three retinal pigment epithelium layers can be visible and evident, or appear fused together.

Retinal drüsen generate irregularities and waves on the retinal pigment epithelium complex and the Bruch's membrane can be easily seen as a thin hyperreflective line.

The retinal pigment epithelium serous detachments deform the retinal posterior limit and forms with the choriocapillaris an angle greater than 45 degrees, whereas the serous detachments of neuroepithelium protrude less and form angles equal to or less than 30 degrees with the retinal pigment epithelium. The Bruch's membrane is visible **in case of RPE detachment (Fig. 8)**.

Retinal Profile Outline

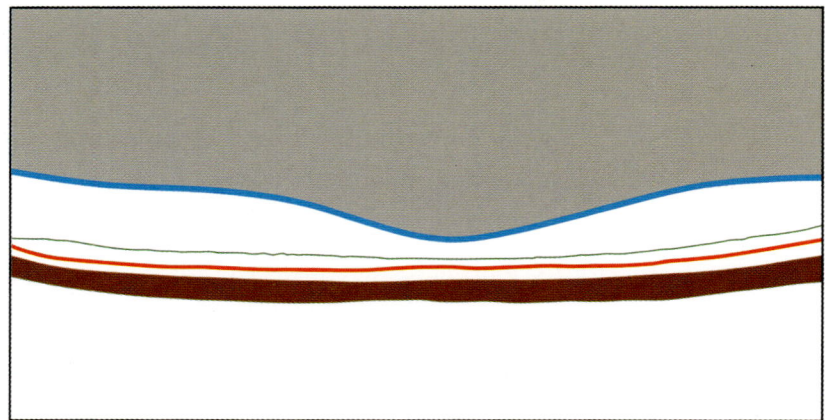

1. Normal profile.

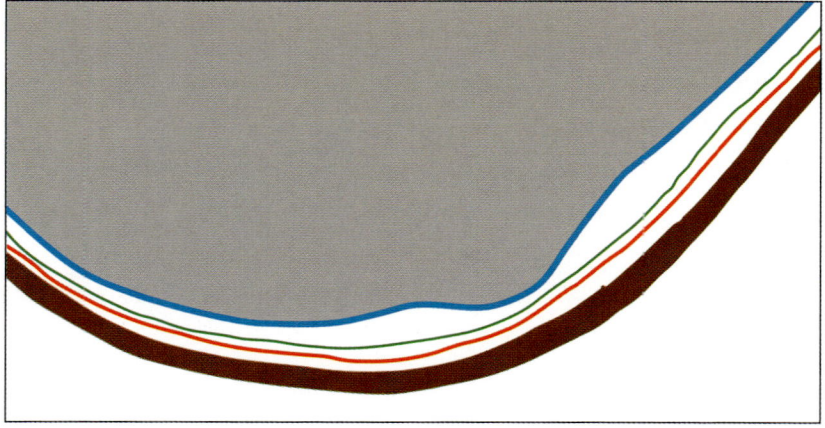

2. Full thickness deformation of the central retina; concavity in a case of strong myopia.

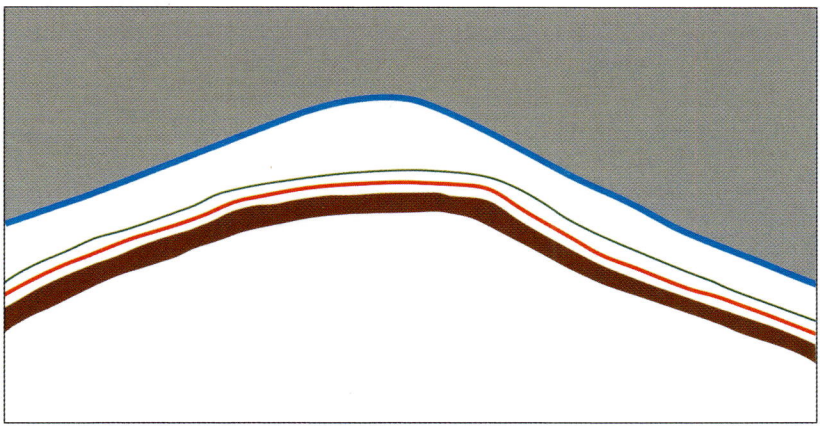

3. Full thickness deformation of the central retina; convexity is dome-shaped in a case of subretinal benign neoformation.

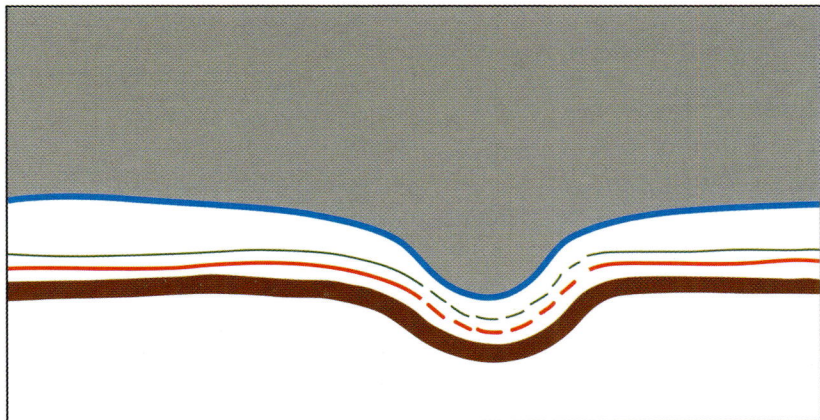

4. Full thickness retinal deformation; localized concavity in case of scleral staphyloma, as a consequence of scleritis.

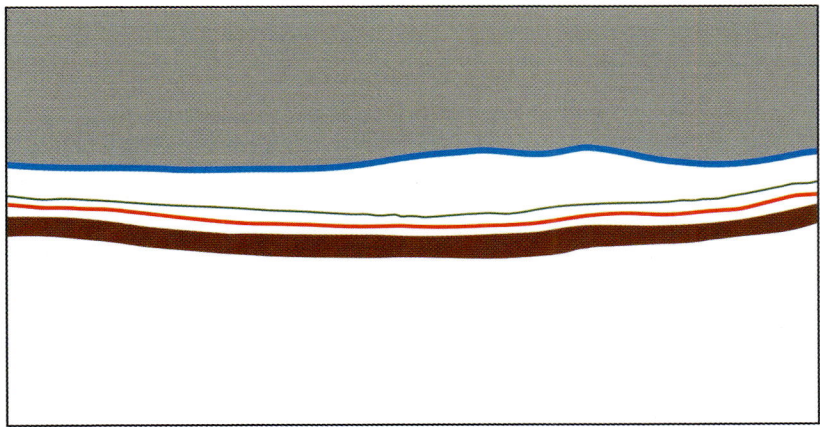

5. Foveal depression is not visible.

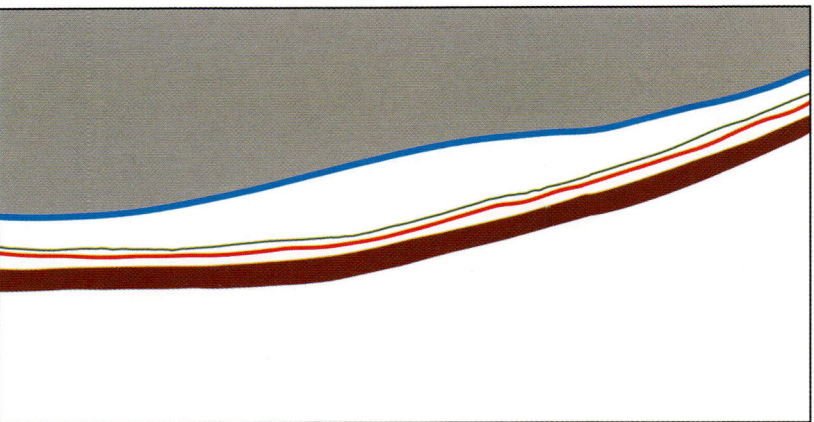

6. Convexity caused by diffuse retinal edema.

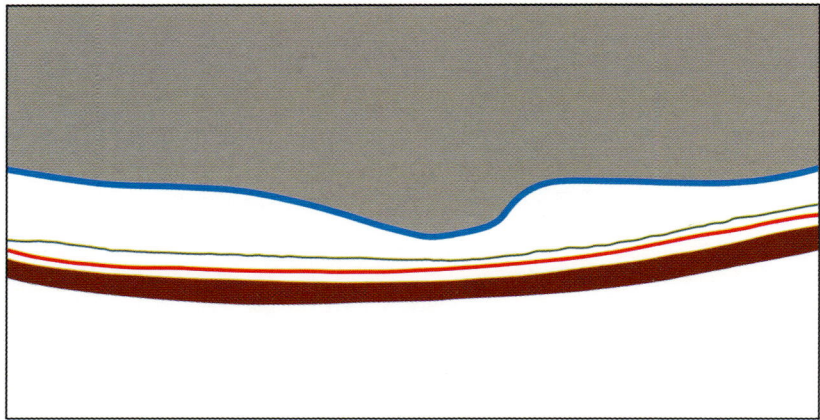

7. Asymmetrical foveal depression in localized edema.

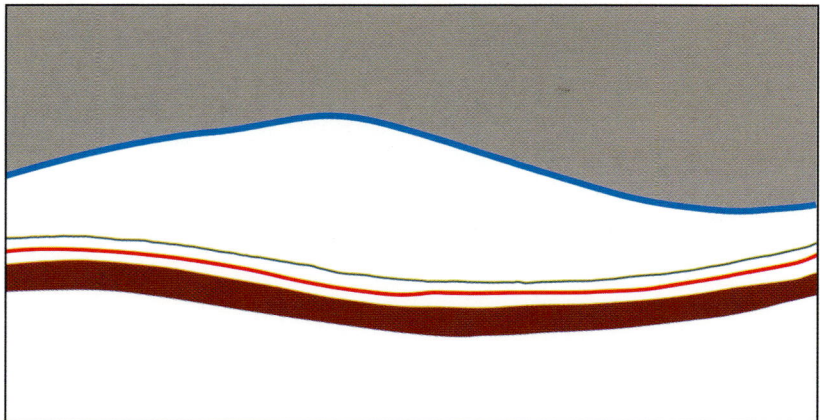

8. Marked retinal thickening with disappearance of foveal depression in a case of diffuse edema in diabetic retinopathy.

QUANTITATIVE ANALYSIS OF PATHOLOGIC OCTS

9. Convexity caused by retinal edema in a case of subretinal revascularization.

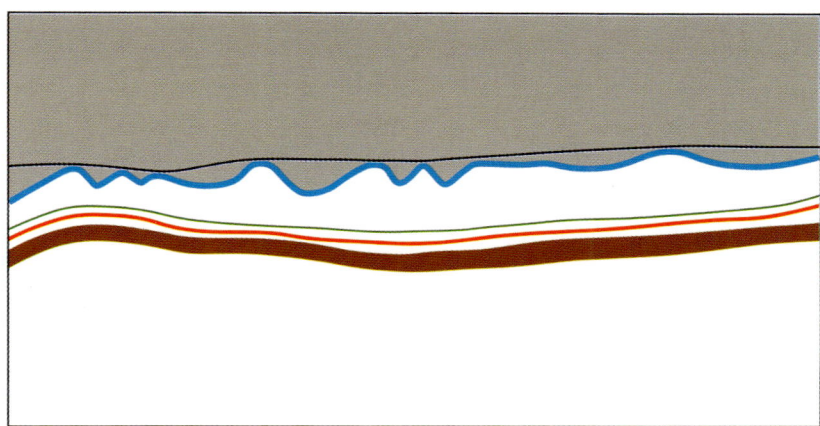

10. Retinal folds caused by horizontal traction exerted by an epiretinal membrane.

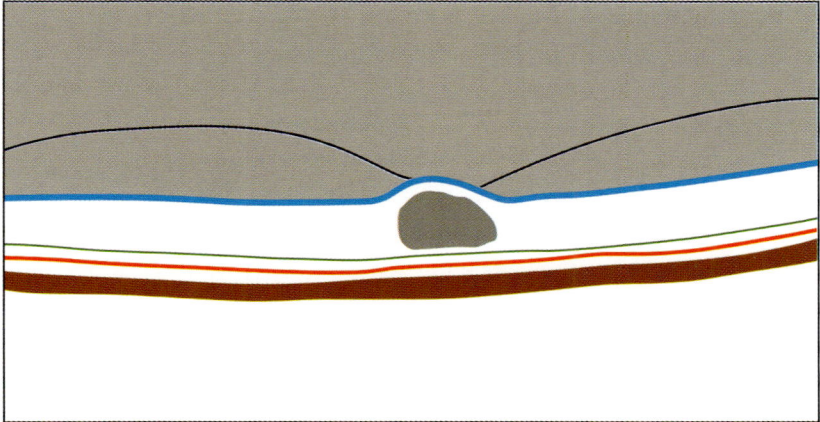

11. Retinal profile deformed by vitreoretinal traction, appearance of impending hole.

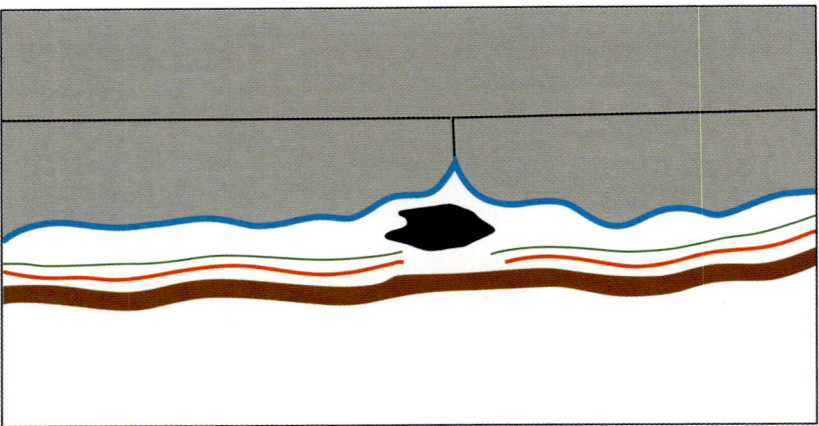

12. Vitreoretinal anteroposterior traction. Retinal profile deformed by traction. Impending hole.

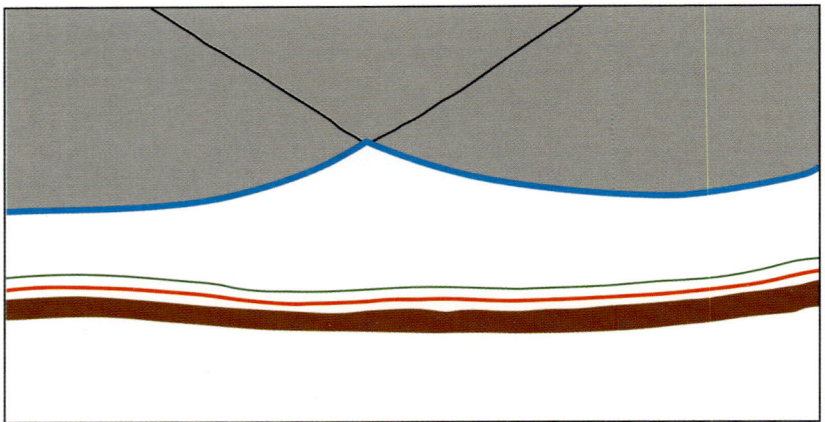

13. Retinal profile deformed by vitreoretinal traction.

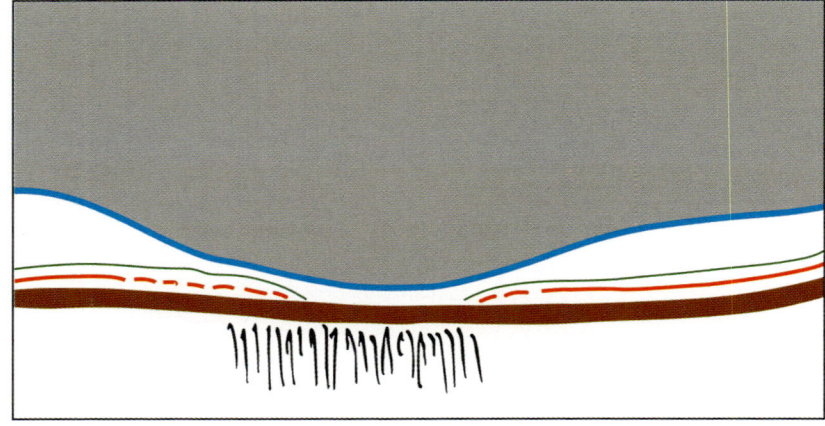

14. Widening and deepening foveal depression. Marked reduction of retinal thickness, alterations or disappearance of external limiting membrane and inner/outer photoreceptor segment junction. Deep penetration of light rays.

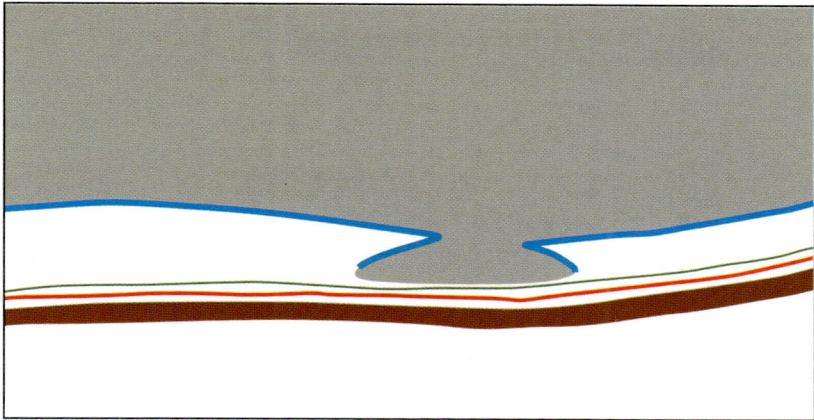

15. Lamellar hole, partial loss of retinal tissue; the posterior retinal layers, the internal limiting membrane and the inner/outer photoreceptor segment junction are intact.

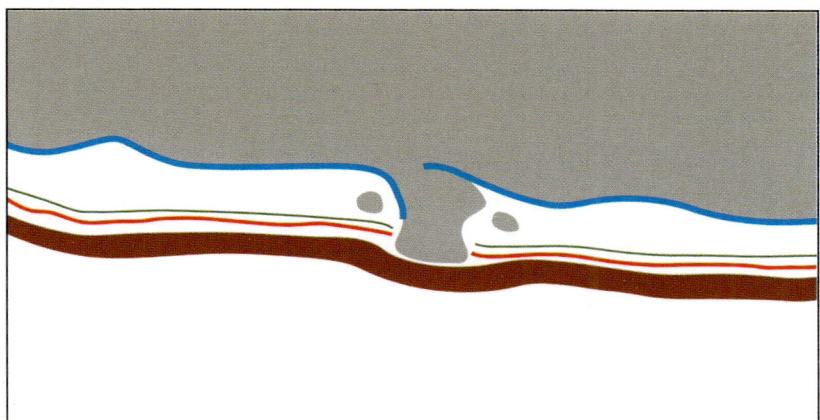

16. Hole with operculum, partial loss of central retinal tissue; the operculum is visible on the hole margin.

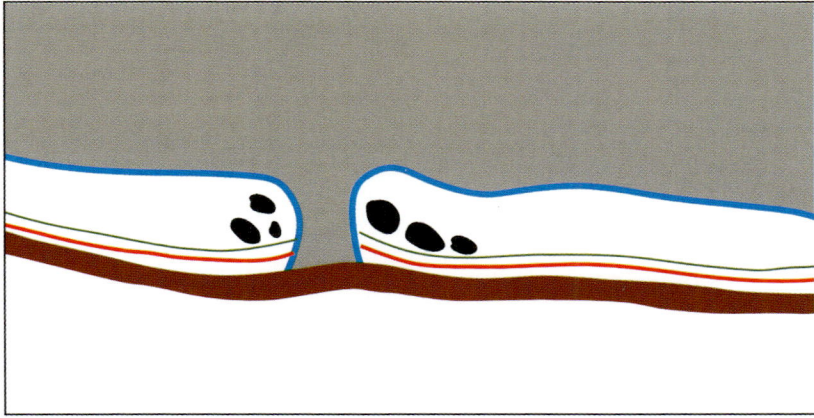

17. Full thickness retinal hole (stage IV), full thickness loss of central retinal tissue, and cystoid edema formation around the hole.

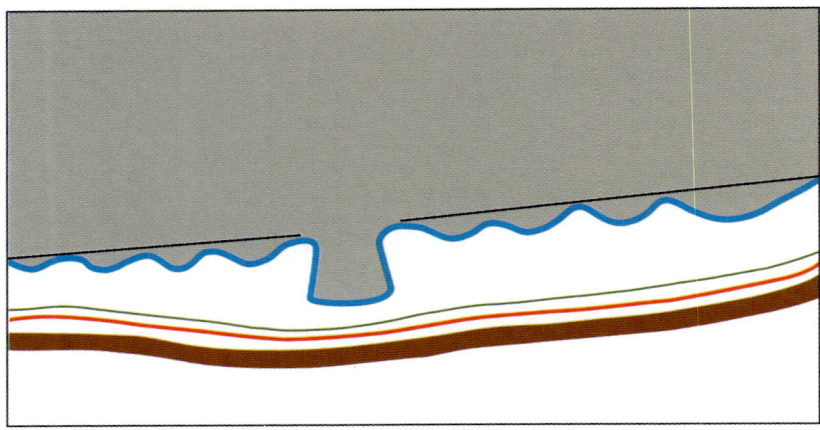

18. Retinal pseudohole. This is not a true retinal hole; there is no tissue loss; retinal tissues are preserved and deformed by tangential tractions exerted by an epiretinal membrane on the retina.

Outer Retina Outline

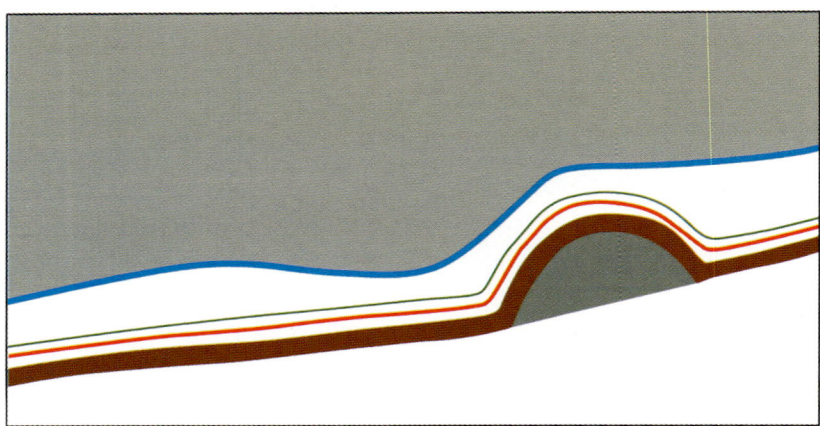

19. Retinal pigment epithelium detachment. The elevation forms an angle greater than 45 degrees with the Bruch's membrane. The external limiting membrane and the inner/outer photoreceptor segment junction are preserved.

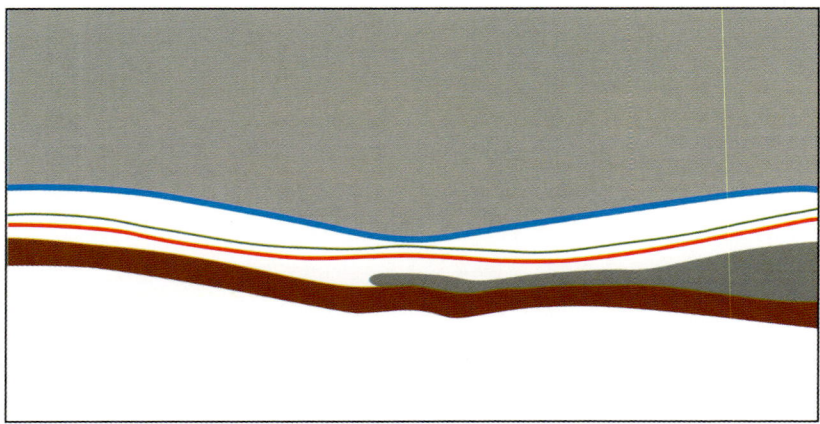

20. Serous detachment of neuroepithelium. The neuroepithelium elevation forms a lesser than 30-degree angle with the retinal pigment epithelium.

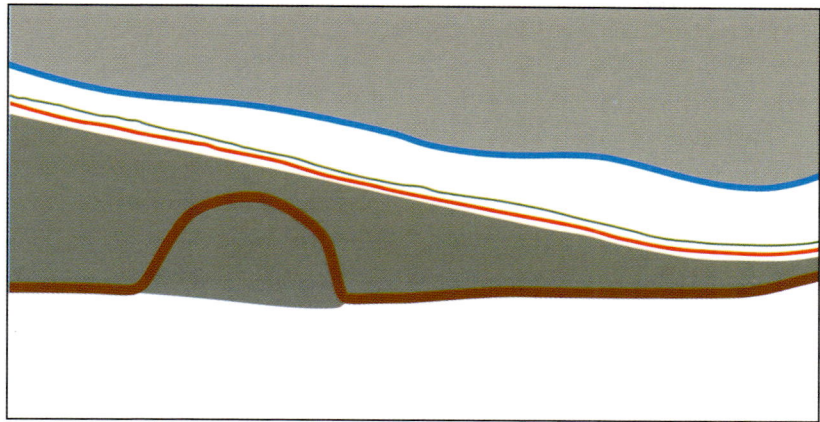

21. Serous detachment of neuroepithelium associated with retinal pigment epithelium detachment. Note how the retinal pigment epithelium elevation forms a greater than 45-degree angle with Bruch's membrane, whereas the neuroepithelium elevation forms a lesser than 30-degree angle with the retinal pigment epithelium.

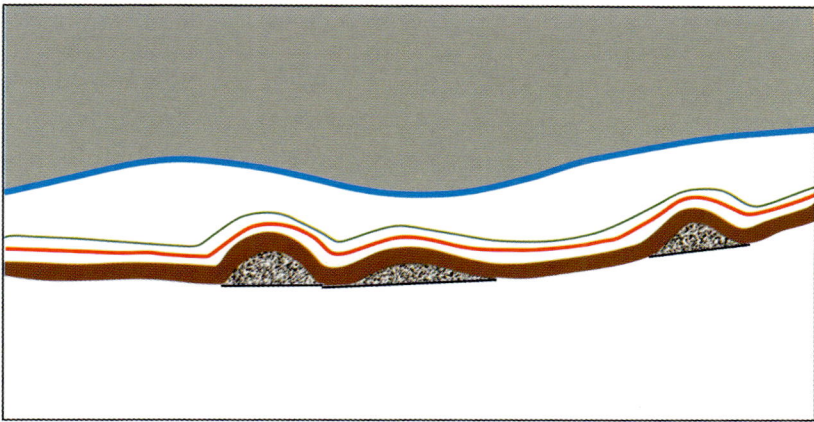

22. Retinal pigment epithelium drüsen: Three drüsen of the retinal pigment epithelium can be seen. Their contents are slightly hyperreflective. Bruch's membrane is also visible as a thin line; the external limiting membrane and the ellipsoid (inner/outer photoreceptor segment junction) are preserved.

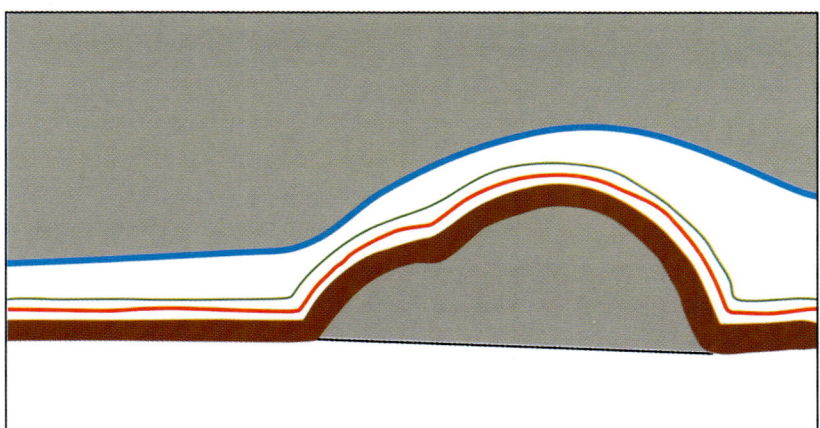

23. Drüsenoid detachment of retinal pigment epithelium. The irregular retinal pigment epithelium elevation lifts the retina as well, deforming its profile. The external limiting membrane and the ellipsoid (inner/outer photoreceptor segment junction) are preserved. The detachment contents are slightly hyperreflective; Bruch's membrane is visible as a thin line.

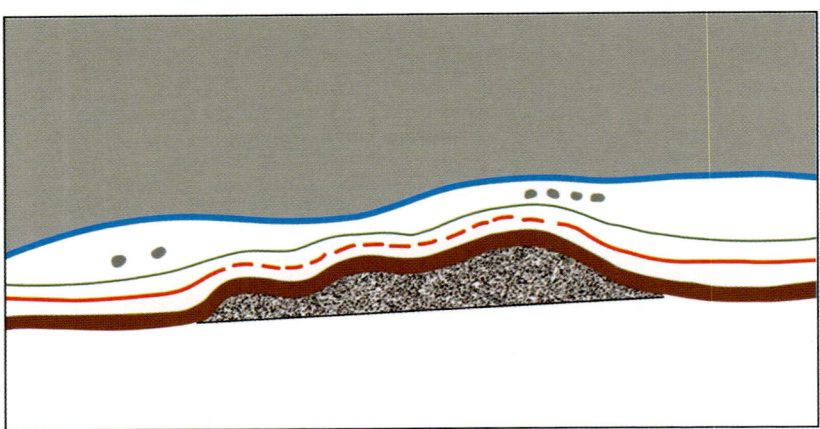

24. Retinal pigment epithelium elevation in macular degeneration with occult neovascularization. Slightly irregular retinal profile; irregular detachment of retinal pigment epithelium with non-homogeneous contents. Bruch's membrane is visible. The retinal pigment epithelium, the ellipsoid (inner/outer photoreceptor segment junction) and the external limiting membrane are altered. Rare cystoid edema cavities are present.

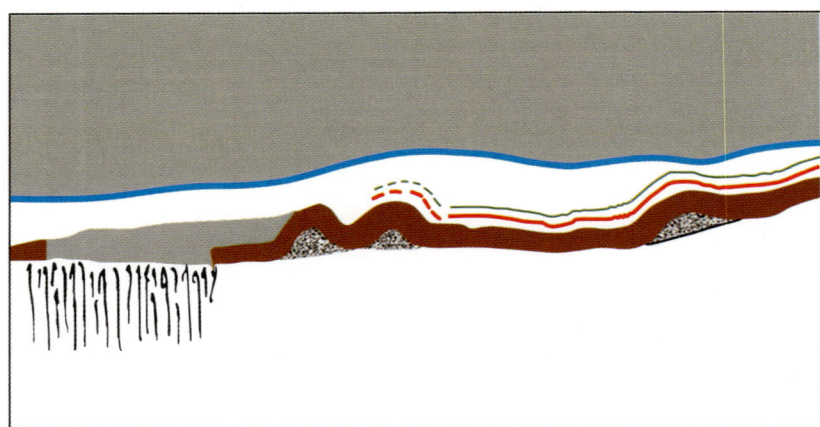

25. Retinal pigment epithelium tear: Short interruption of the retinal pigment epithelium, where, lacking a pigmented screen, the OCT shows a deep light penetration. Retraction of residual retinal pigment epithelium forming folds, localized serous neoepithelium elevation, alterations of the external limiting membrane and of the ellipsoid (inner/outer photoreceptor segment junction).

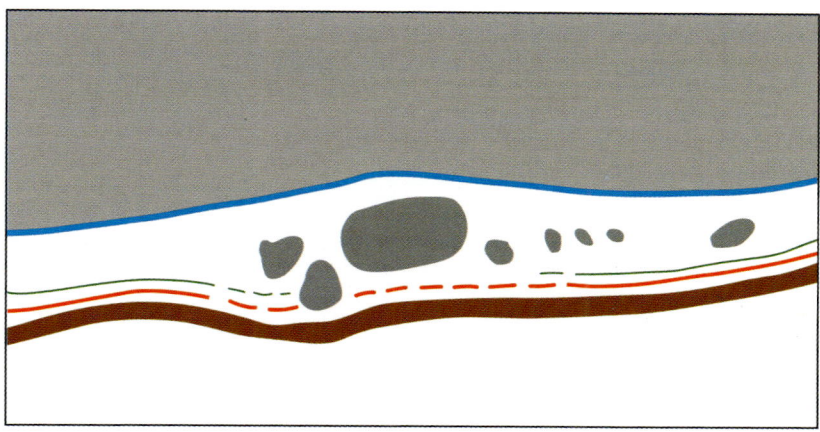

26. Macular cystoid edema in diabetic retinopathy. Presence of irregular cavities of small to large cystoid edema cells and deep hard exudates, with localized lesions of photoreceptors, external limiting membrane and ellipsoid (inner/outer photoreceptor segment junction).

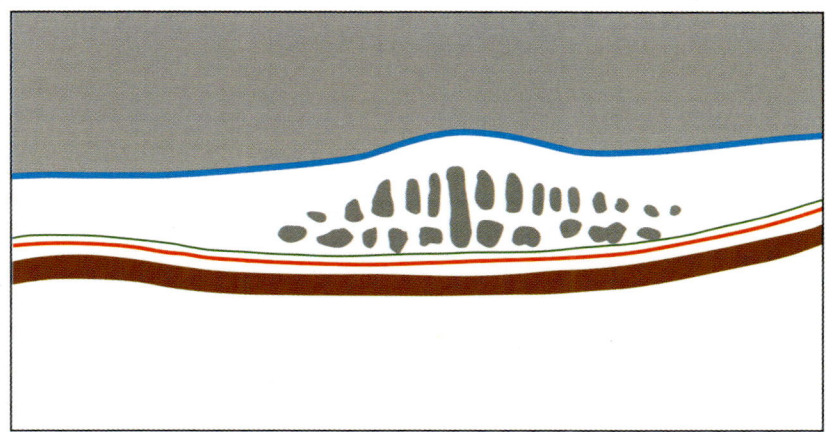

27. Macular cystoid edema, increased retinal thickness with retinal profile deformation. In this case of edema following cataract surgery, we can see the presence of small regular edema cells within the inner and outer nuclear layers. The external limiting membrane, the ellipsoid (inner/outer photoreceptor segment junction) and the retinal pigment epithelium are normal.

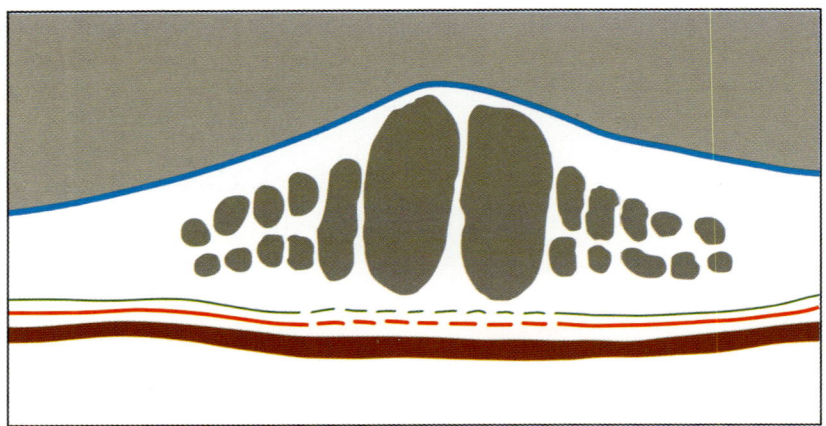

28. Advanced macular cystoid edema. Retinal profile deformation and increase in retinal thickness. Two regular rows of cystoid edema cavities can be seen at the level of the inner and outer nuclear layers. Within the fovea, some pseudocysts merge vertically, forming ovular cells with a vertical axis. There are alterations of the external limiting membrane and the ellipsoid (inner/outer photoreceptor segment junction). Retinal pigment epithelium is normal.

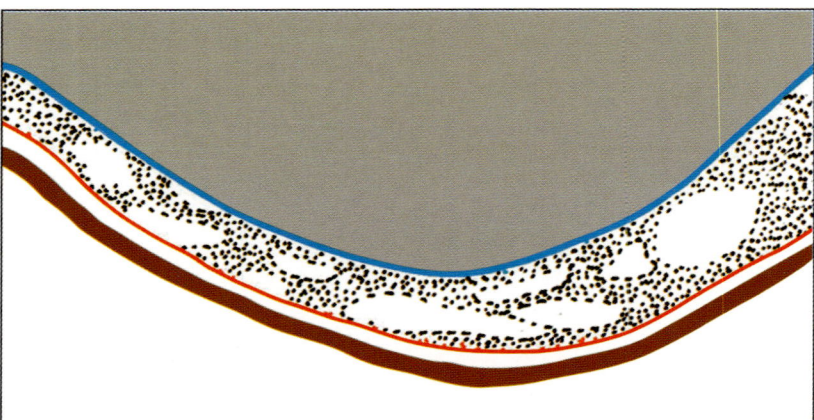

29. Myopic retinoschisis: The retinal tissue has been delaminated, but preserves numerous shoots that follow the horizontal layout of the retinal layers. Retinoschisis cavities appear elongated and roughly parallel to the retinal surface and the retinal pigment epithelium. These cavities are angular and the delaminated tissue layers tend to form sharp angles with each other.

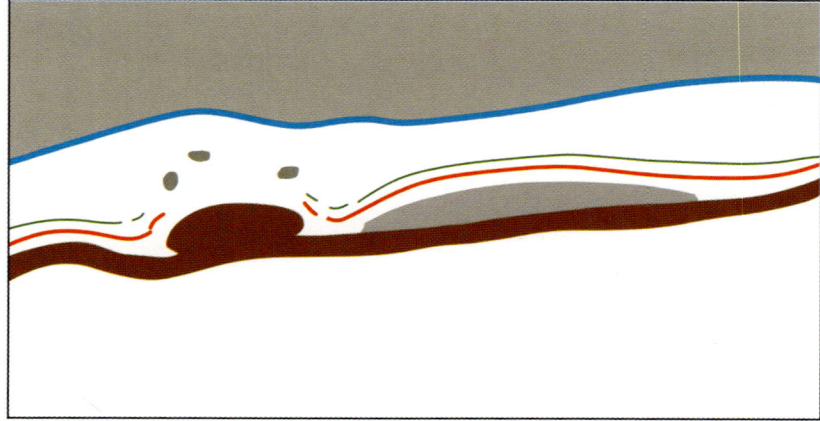

30. Low neuroepithelium detachment, few cystoid edema cavities close to a classical neovascular membrane. Alterations of the external limiting membrane and the ellipsoid (inner/outer photoreceptor segment junction).

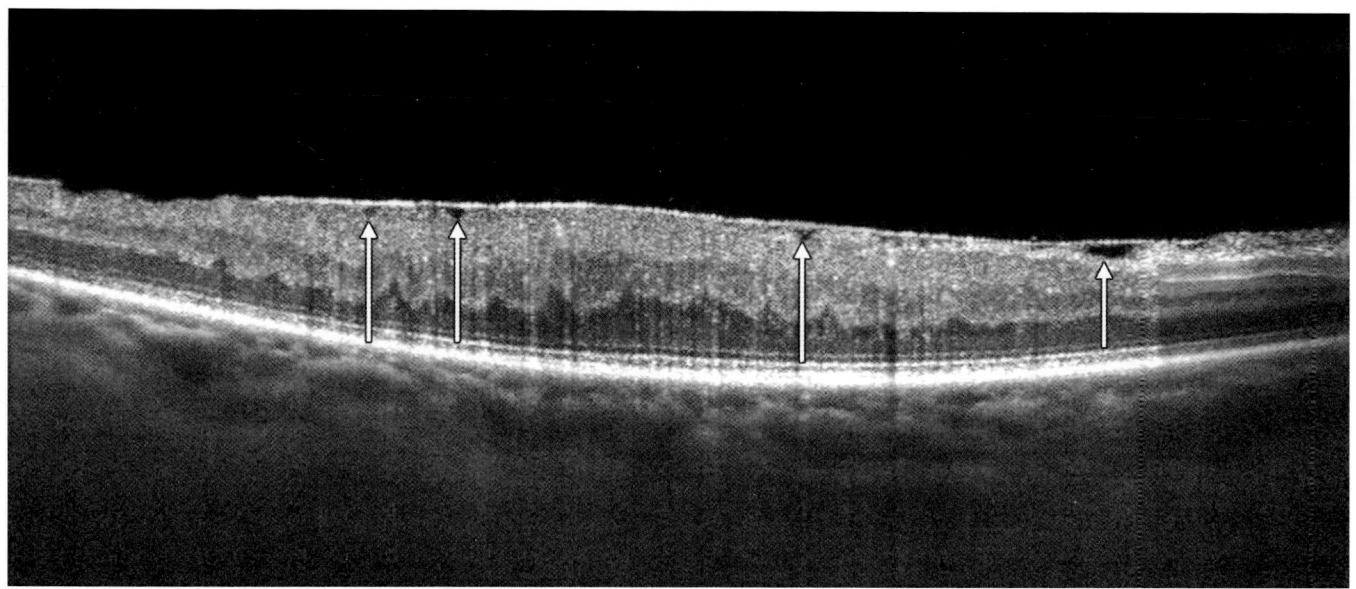

FIGURE 1: ADHERENT EPIRETINAL MEMBRANES: MORPHOLOGIC ALTERATIONS – RETINAL DEFORMATION – LOSS OF NORMAL PROFILE

Deformation of the retinal surface, disappearance of the foveal depression, diffuse retinal edema, presence of adherent epiretinal membranes. These adhesions form rare small folds on the retinal surface. Arrows show retinal folds. The outer nuclear layer, which contains the photoreceptor nuclei, presents a diffuse edema; the external limiting membrane, the ellipsoid (inner/outer photoreceptor segment junction) and the retinal pigment epithelium/choriocapillaris complex are normal. The choroid is normal.

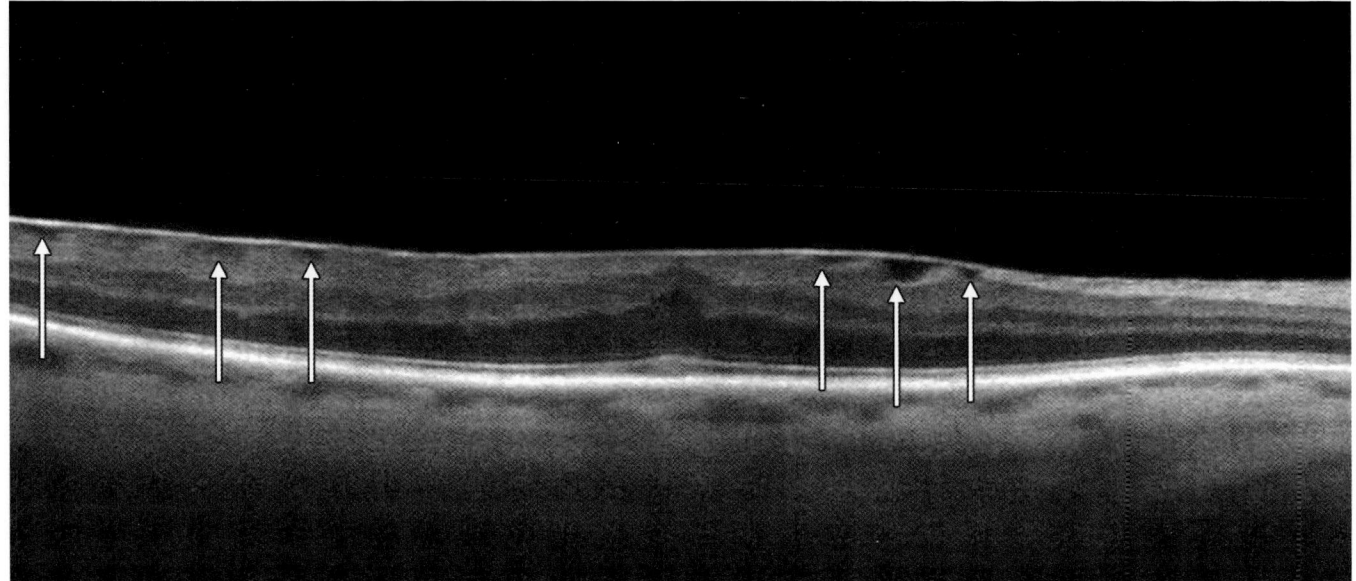

FIGURE 2: MACULAR PUCKER – MORPHOLOGIC ALTERATIONS – RETINAL DEFORMATION – LOSS OF NORMAL PROFILE

The retina has lost its normal profile, appears irregular, deformed by folds caused by a very adherent epiretinal membrane, with transversal traction. Arrows show retinal folds.

Disappearance of foveal depression. The retinal thickness is increased by a diffuse edema; the external limiting membrane and the ellipsoid (inner/outer photoreceptor segment junction) are normal.

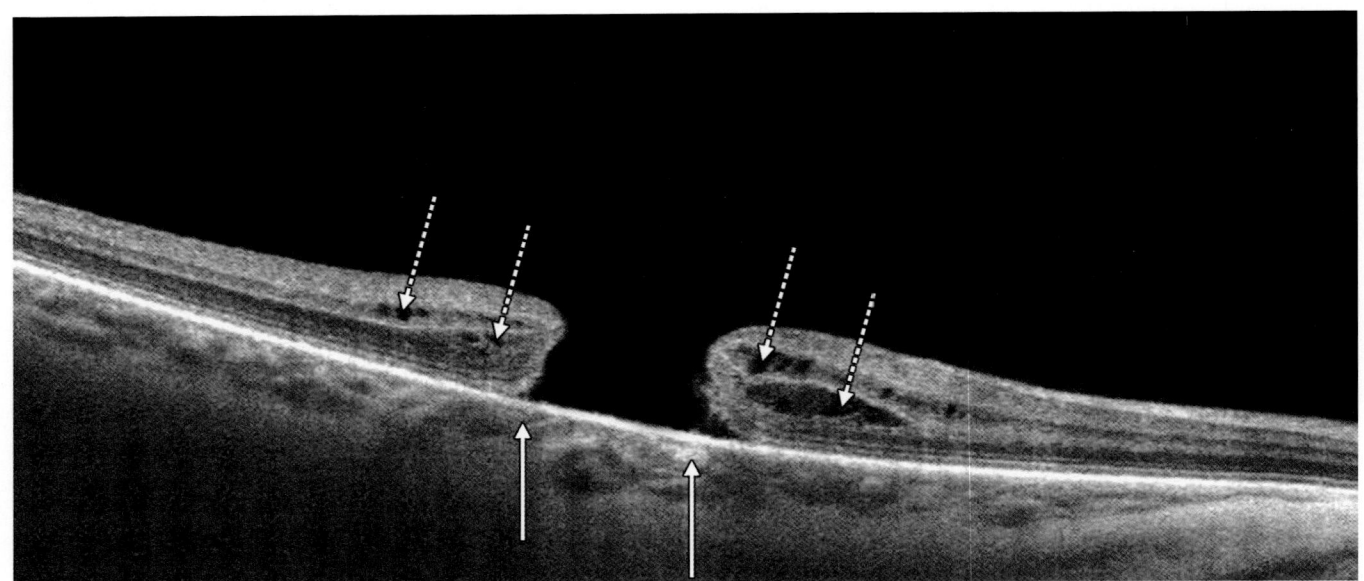

FIGURE 3: FULL THICKNESS MACULAR HOLE – MORPHOLOGIC ALTERATIONS – LOSS OF NORMAL PROFILE

The retina has lost its normal profile and presents a very significant tissue loss in all retinal layers. The retinal thickness has increased at the hole margins, with fluid in the intraretinal pseudocysts and the cystoid edema small cavities surrounding the macular hole within the inner and outer nuclear layers. The outer nuclear layer, which contains the photoreceptor nuclei, the external limiting membrane and the ellipsoid (inner/outer photoreceptor segment junction) are interrupted, thus explaining the serious loss of vision. Line arrows show the limits of the macular hole, dashed arrows show the intraretinal pseudocysts.

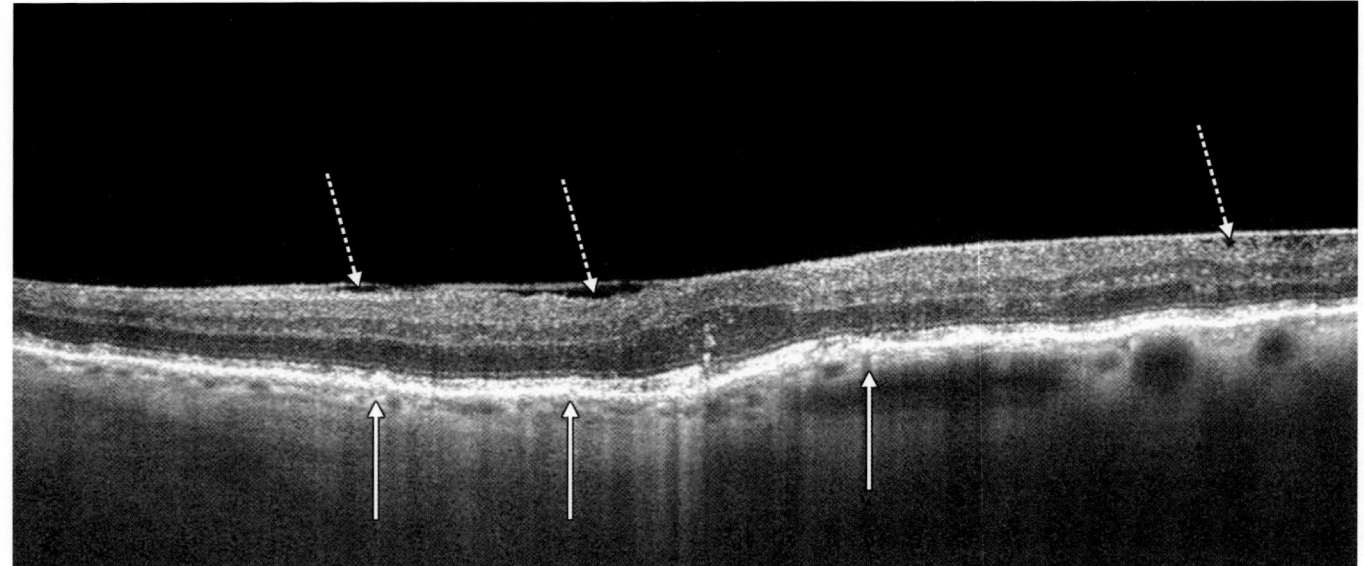

FIGURE 4: EPIRETINAL MEMBRANE – MORPHOLOGIC ALTERATIONS – LOSS OF NORMAL PROFILE

The retina has lost its normal profile. It appears irregular and deformed by the transversal traction of an epiretinal membrane that is very adherent and causes rare retinal folds. The retinal thickness is slightly increased by diffuse edema at the nuclear layer level: the external limiting membrane is almost normal, whereas the ellipsoid (inner/outer photoreceptor segment junction) is irregular, with altered thickness and some localized interruptions. The retinal pigment epithelium appears slightly irregular and allows light rays to penetrate in the choroid that is very thin and atrophic (Left hand half of the figure). On the right figure half, the choroid appears normal. Line arrows show the pigment epithelium lesions, dashed arrows show the retinal folds.

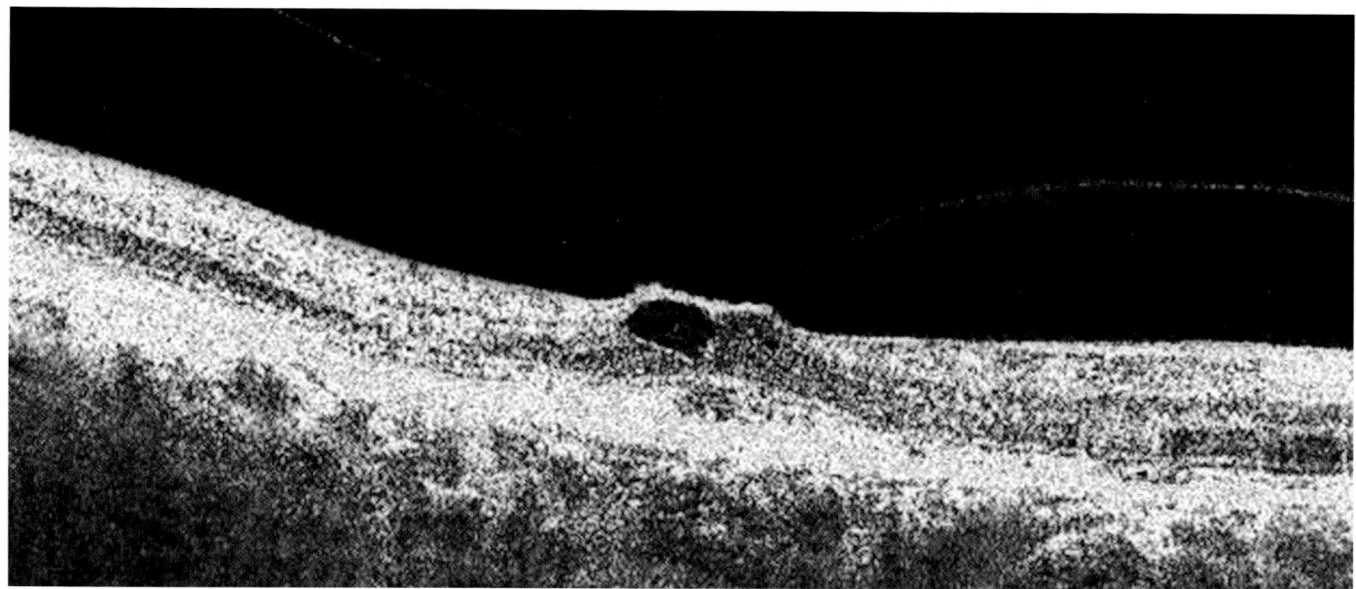

FIGURE 5: VITREORETINAL ANTEROPOSTERIOR TRACTION – PROFILE LOSS – STRUCTURAL ALTERATIONS
Membranes adhere to the retinal surface, with strong anteroposterior traction. This traction has caused a localized retinal elevation and the formation of a single trapezoidal cavity under the retinal surface.

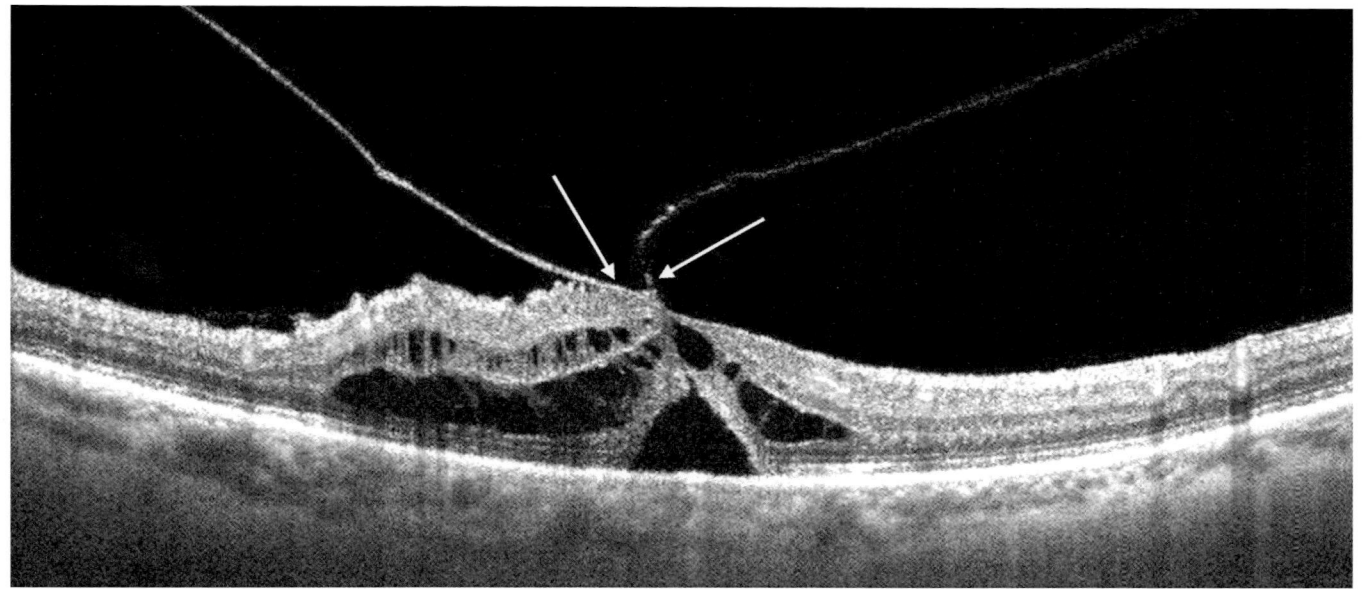

FIGURE 6: VITREORETINAL ANTEROPOSTERIOR TRACTION – HOLE IN PROGRESS – PROFILE LOSS – STRUCTURAL ALTERATIONS
Abrupt onset of vitreoretinal traction in a 60-year-old subject, whose OCT, taken two weeks before, was normal. The vision has been reduced from 20/20 to 20/40. Multiple intraretinal cavities can be seen, the most important of which is under the retinal surface. This cavity has an irregular, anvil-like shape. Around the cavity of the hole in progress small slits are seen, mainly at the inner nuclear layer level. Slit shape is determined by location within the parallel retinal layers.

Two membranes adhere to the retinal surface and exert strong anteroposterior traction. Arrows show the vitreo-retinal traction.

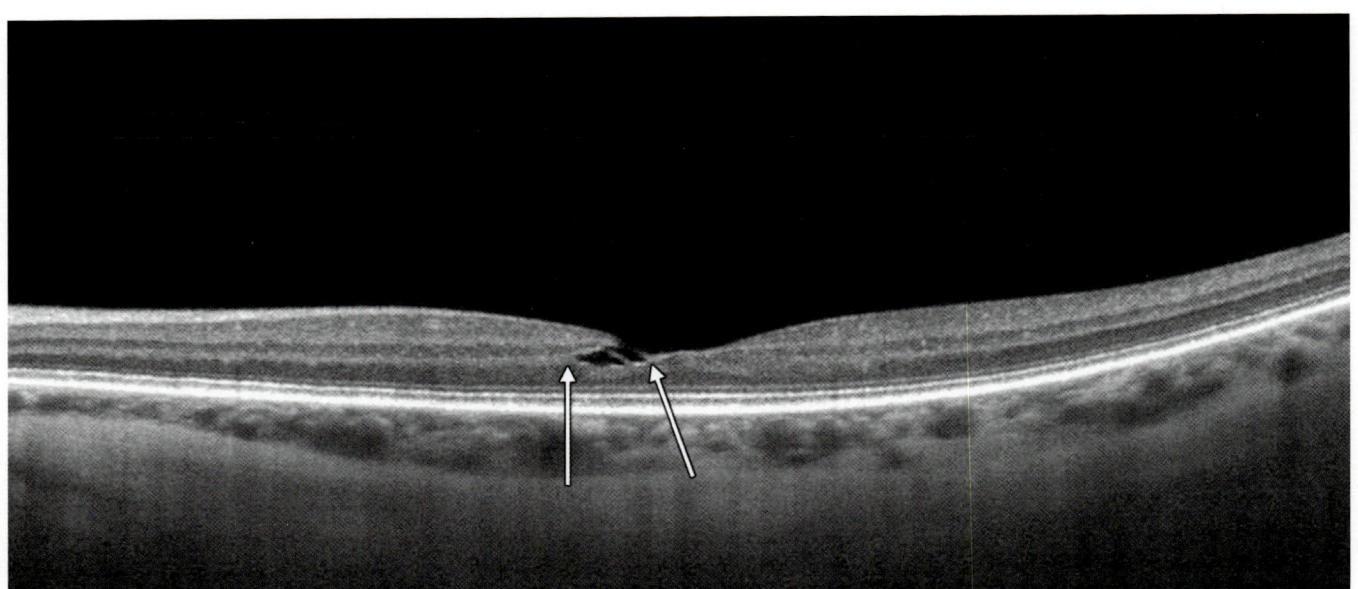

FIGURE 7: ASYMMETRIC LAMELLAR HOLE – MORPHOLOGIC ALTERATIONS – LOSS OF NORMAL PROFILE
The retina has lost its normal profile in correspondence to the retinal inner layers. An asymmetric horizontal slit can be observed at the outer plexiform layer level. Optically empty, minimal cavities, associated to cystoid edema, can be seen at the level of the nuclear layers. The slit shape is determined by its position between parallel retinal layers. The outer nuclear layer, which contains the photoreceptor nuclei, the external limiting membrane, the ellipsoid (inner/outer photoreceptor segment junction) and the retinal pigment epithelium/choriocapillaris complex are preserved, thus explaining the patient's good residual sight. Line arrows show the limits of the assymmetric lamellar macular hole.

TABLE 2 Morphologic alterations

- **Global retina deformations:**
  - Myopic concavity
  - Convexity: presence of a choroidal tumor
  - Scleral staphyloma
  - Dome-shaped macula
- **Retinal profile alterations**
- **Structural intraretinal alterations**
- **Posterior layers deformation:**
  - External limiting membrane
  - Ellipsoid (inner/outer photoreceptor segment junction)
  - Retinal pigment epithelium
  - Bruch's membrane.

QUANTITATIVE ANALYSIS OF PATHOLOGIC OCTS

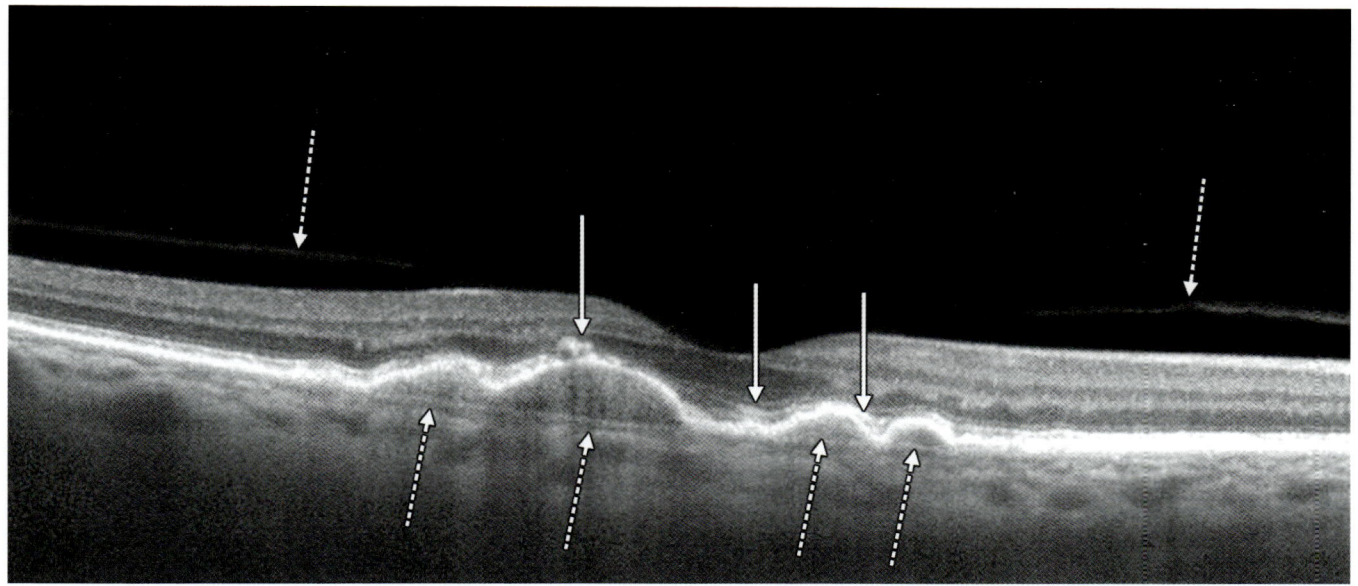

FIGURE 8: DRÜSEN – DEEP MORPHOLOGIC ALTERATIONS – HIGH REFLECTIVITY
In this case of early AMD, the drüsen are seen as wavelets in the retinal pigment epithelium, with average reflectivity. The outer nuclear layer, which contains the photoreceptor nuclei, has a slightly altered thickness; the external limiting membrane and the ellipsoid (inner/outer photoreceptor segment junction) follow the wavy layout and show slight alterations. At fovea level, limiting membrane and ellipsoid are broken up at one point. Bruch's membrane appears as a thin, highly reflective, horizontal line in correspondence to the retinal pigment epithelium elevation. The choroid thickness is decreased. Line arrow shows the limiting membrane, junction of the inner/outer photoreceptors segments pigment and epithelium lesions, dashed arrow shows Bruch's membrane. The dashed arrow in the vitreous shows detached hyaloid.

Choroid

The choroid vessels are hyporeflective and can be differentiated in two layers: the Sattler's layer of small vessels and the Haller's layer of large vessels. Each layer can present a decreased or increased thickness according to pathology, age and refraction. The connective tissue between vessels shows density and visibility variations. Externally the lamina fusca appears as a thin dark line. Usually, the suprachoroidal space is not visible, except in certain pathologies **(Figs 9 and 10)**.

TABLE 3	Retinal profile cross-section scan

- Normal profile
- Increased foveal depression
- Reduced foveal depression
- Asymmetric foveal depression
- Absence of depression
- Dome-shaped foveal area
- Retinal convexity caused by edema or traction, (instead of depression)
- Profile deformed by localized vitreoretinal traction
- Profile deformed by complex vitreoretinal tractions
- Retinal profile deformed by wavy retinal folds
- Retinal profile wrinkled by tight retinal folds.

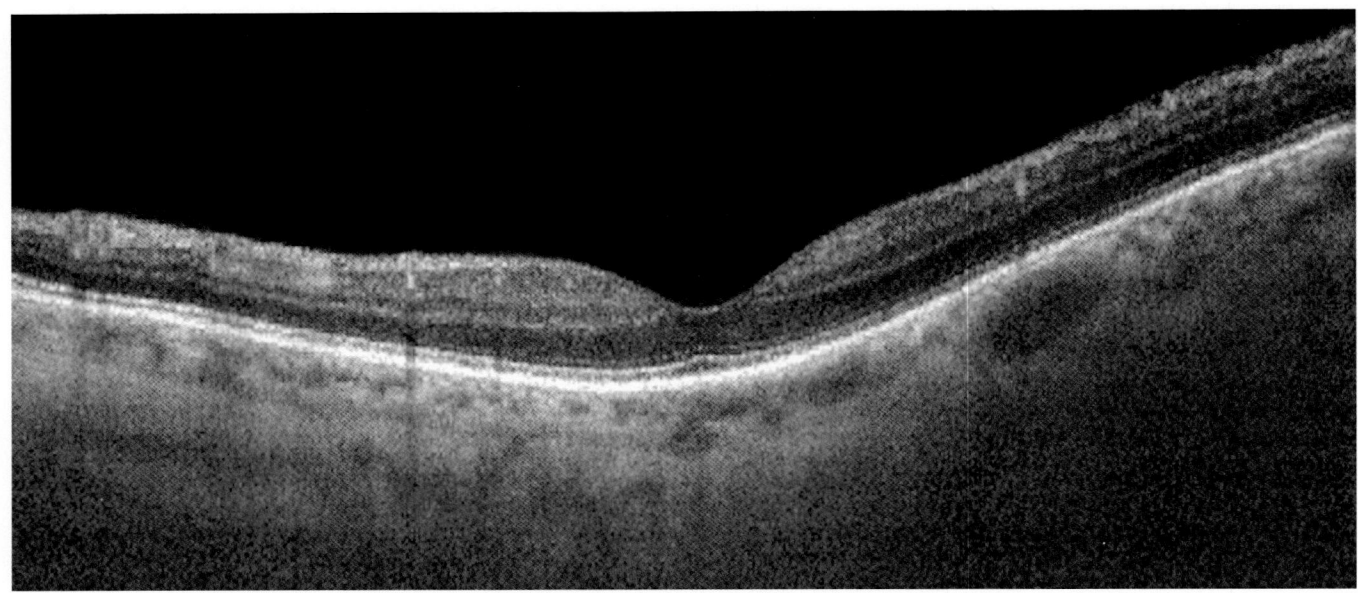

FIGURE 9: NORMAL CHOROID
The cross-section scans show a normal retina, and all layers are well seen; choroid is thin, about 150 micron in thickness at the fovea; equally visible are Sattler's small vessels and Haller's larger ones. A thin dark line is visible along the demarcation between choroid and sclera, representing the lamina fusca and the virtual suprachoroidal space. Below the fovea we can discriminate the sclera with oblique intrascleral channels. The interstitial tissue presents normal reflectivity.

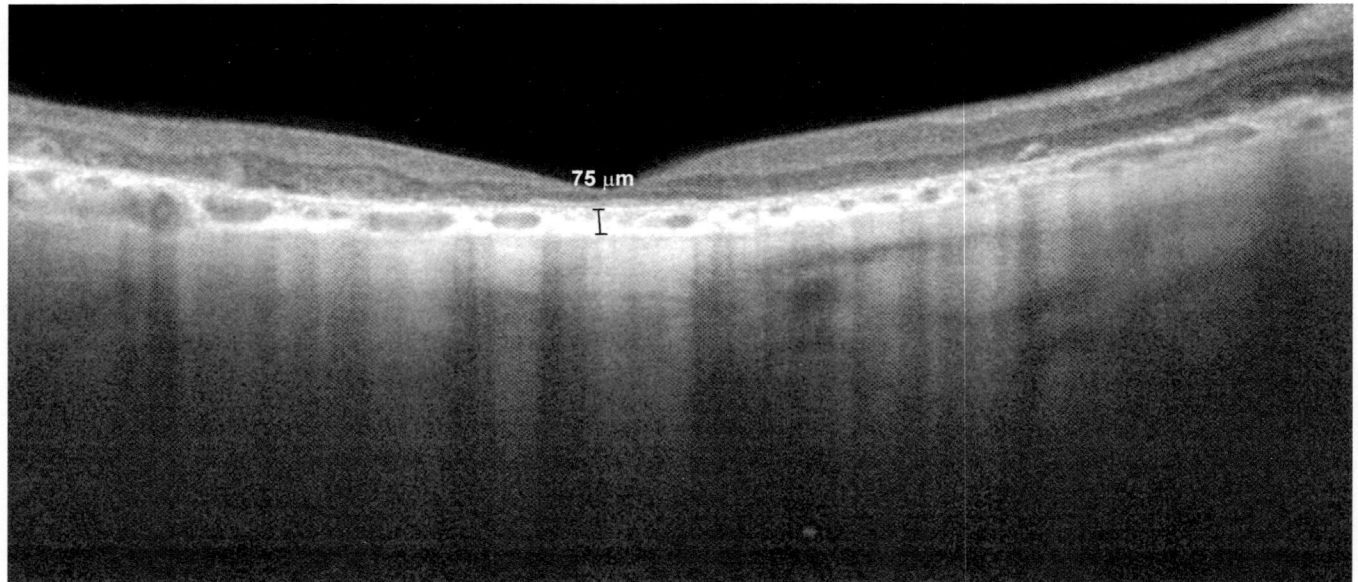

FIGURE 10: SUBATROPHIC CHOROID
Linear measurement of the choroid thickness in a case of atrophic macular degeneration. The caliper has been set at the external margin of the retinal pigment epithelium and at the external boundary of the visible choroid. The wall of the choroid vessels is hyperreflective. The choroid thickness is greatly reduced (75 μm).

Retinal Structure Study: Segmentation

Structure analysis studies the interrelations and layout of the parts within a complex entity, in our case the retina. Segmentation studies the layout and the arrangement of tissue forming a regular stratified structure in the retina.

The retina consists of a complex stratified formation of cellular layers, fibers and capillaries organized in structures that effectively form horizontal barriers.

The retina infrastructure is supported by horizontal and vertical formations.

The vertical (cell chains, Müller cells) and horizontal formations explain the position, dimension and shape of exudates, hemorrhages and cystoid cavities, which we can analyze using OCTs. Vertical and horizontal anatomic barriers block the spreading of pathologic processes.

The retinal inner structure appears on OCTs as bands of homogeneous reflectivity, both high and low.

Initially, each retinal layer must be considered by itself. Subsequently, if anomalies emerge, we study the relationship between the individual layers.

The OCT image segmentation first study separately a group of retinal layers; then the study recombines together these layers and reconstructs the OCT image itself. The segmentation technique is an important progress in the study of retinal alterations. The first segmentation automatically performed by all the OTC instruments has been the identification of the nerve fiber layer. In fact, the study of the nerve fiber thickness is essential to the study of glaucoma and its follow-up. Ganglion cell segmentation is important for the study of glaucoma, Parkinson disease, Alzheimer disease and other neurodegenerative disorders.

STRUCTURE: SEGMENTATION

OCT imaging allows performing the segmentation of the chorioretinal layers both in 2D and 3D images. Due to the fact that retina and choroid form a regular stratified structure and retina components constitute stacked layers with diverse reflectivity, the segmentation techniques can highlight the stratification of the eye membranes. Each layer is separately identified based upon its reflectivity.

OCT devices allow acquiring a series of cross-section scans that form a cube of tissue to be examined from various sides and to be dissected. Then, the software subtracts some layers and highlights others layers, especially the nerve fibers, inner and outer retina, retinal pigment epithelium and choroid. It allows highlighting of the internal limiting membrane sheet, the retinal pigment epithelium sheet and a block that reconstructs the retina, from the internal limiting membrane to the retinal pigment epithelium.

Thus, 3D images of the retina surface, the retinal pigment epithelium or other layers can be acquired. These 3D images are very useful for teaching purposes.

INNER RETINA AND OUTER RETINA

OCT instruments may provide a quick and easy segmentation of the inner and outer retina.

The inner retina includes the nerve fiber layer, formed by ganglion cell axons, ganglion cell bodies, ganglion layer and inner plexiform layer, where the ganglion cells are articulated with the bipolar cells. These three layers form the ganglion cell complex, which becomes thinner when the ganglion cells die following the glaucoma evolution. This layer is also very sensitive to lesions caused by retinal ischemia and to the occlusion of retinal arteries. Inner retina is also affected in neurodegenerative disorders as Alzheimer and Parkinson diseases.

The outer retina is formed by the layers between retinal pigment epithelium and inner plexiform layer, and thus includes the ellipsoid (inner/outer photoreceptor segment junction), the outer nuclear layer composed of photoreceptor nuclei, the outer plexiform layer, and the inner nuclear layer formed by bipolar cell nuclei.

The segmentation allows measuring the inner and outer retina thickness and volume. Furthermore, it allows studying the localized alterations of retinal segments, analyzed through maps or 3D images.

Soon new softwares will be able to segment with precision each retinal layer, highlighting retinal membranes, junctions and articulations. The separate segmented study of certain structures is already possible and important.

Nevertheless, at times, the segmentation cannot be correctly performed. The regular stratified structure is modified by retinal pathology and sometimes disarranged. For instance, in case of advanced edema, the segmentation data are not reliable. While the segmentation of the normal retina can be performed with almost perfect precision, in the presence of irregularities, edemas and relevant atrophies of retinal layers, the segmentation lines become inaccurate and often the instrument fails to follow the limits of a given retinal layer.

Currently, the segmentation is useful to the study of moderate retinal alterations. For example, in case of acute ephithelitis, multiple evanescent white dots syndrome (MEWDS), acute zonal occult outer retinopathy (AZOOR) or Tamoxiphen poisoning, the ellipsoid (inner/outer photoreceptor segment junction) can be faded, unfocused, interrupted in certain points, dashed, invisible or absent **(Figs 11 and 12)**.

Congenital and acquired retinoschisis present spontaneous separations of retinal layers and offer a good example of pathologic segmentation **(Tables 4 to 17)**.

TABLE 4	Causes of retinoschisis
<ul><li>Macular juvenile X-linked retinoschisis</li><li>Acquired peripheral retinoschisis in the adult</li><li>Goldmann-Favre vitreo-tapetoretinal degeneration.</li></ul>Retinoschisis secondary to:<ul><li>Traction</li><li>Retinal slit</li><li>Hole in formation</li><li>High myopia</li><li>Optic pit</li><li>Vitreous retinopathy (Wagner syndrome)</li><li>Post-inflammatory.</li></ul>	

TABLE 5 — Nuclear and plexiform layers

- Evaluation of thickness, reflectivity and possible abnormal formations
- In case of anomalies, the relationship between the single layers' thickness should be studied.

TABLE 6 — Inner nuclear layer

- Normal
- Globally reduced thickness
- Localized reduced thickness
- Reduced thickness in some points
- Increased thickness
- Increased thickness due to cystoid edema pseudocysts
- Presence of exudates in the layer thickness
- Loss of the inner nuclear layer.

TABLE 7 — Outer nuclear layer

- Normal
- Globally reduced thickness
- Localized reduced thickness
- Reduced thickness in some points
- Increase in thickness
- Increased thickness due to cystoid edema
- Presence of exudates or other materials
- Total loss of the outer nuclear layer.

TABLE 8 — External limiting membrane (1)

- Integrity
- Discontinuity
- Reflectivity
- Abnormal structures.

TABLE 9 — External limiting membrane (2)

- Normal linear appearance
- Deformed by retinal pigment epithelium folds and detachments
- Thickened and altered
- Faded, unfocused
- Interrupted in some points—dashed
- Non-visible
- Total loss.

TABLE 10	Ellipsoid (Inner/outer photoreceptor segment junction) (1)

- Integrity
- Discontinuity
- Abnormal structures.

TABLE 11	Ellipsoid (Inner/outer photoreceptor segment junction) (2)

- Normal linear appearance
- Deformed by retinal pigment epithelium folds and detachments
- Wavy appearance (drüsen)
- Wavy appearance, elevation (retinal pigment epithelium detachments)
- Thickened and altered
- Faded, unfocused
- Interrupted in some points—dashed
- Fragmentation
- Disaggregation
- Disorganization
- Non-visible
- Total loss

TABLE 12	Retinal pigment epithelium

- Normal with three visible layers
- Normal, but the layers are not visible, due to their fusion
- Increased thickness
- Reduced thickness
- Irregular thickness due to the disappearance of one or two layers
- Abnormal appearance of layers
- Abnormal reflectivity of one or more layers
- Retinal pigment epithelium fragmentation
- Retinal pigment epithelium disaggregation
- Retinal pigment epithelium disorganization
- Wavy appearance (drüsen)
- Wavy appearance, elevation (Retinal pigment epithelium detachments)
- Presence of abnormal tissues in contact with retinal pigment epithelium: fibrovascular tissue, neovascular tissue, lipofuscin deposits
- Increased reflectivity due to backscattering
- Retinal pigment epithelium deformed by abnormal tissue
- Retinal pigment epithelium deformed by abnormal deposits
- Abnormal structures.

TABLE 13	Bruch's membrane

- Visibility
- Integrity
- Continuity
- Discontinuity.

TABLE 14	Lesions of outer layers and photoreceptors

Lesions of outer layers and photoreceptors are observed in case of:
- Macular degenerations
- Macular retinoschisis
- Acute and chronic retinal epitheliopathy
- Hutchinson Siegrist traumatic epitheliopathy
- Occlusions of branch veins
- Arterial occlusions
- Eclipse retinopathy
- Acute retinitis
- Best's disease
- Acute epithelitis
- Multiple evanescent white dots syndrome (MEWDS)
- Acute zonal occult outer retinopathy (AZOOR)
- Tamoxiphen poisoning
- Retinal ischemia
- Retinal cystoid edema.

TABLE 15	Subretinal deposits

Subretinal deposits are observed in case of:
- Macular degenerations
- Best's disease
- Tamoxiphen poisoning
- Lipofuscin
- Occult neovascularization
- Hemorrhage's outcome.

TABLE 16	Differential diagnosis of isolated lesions of outer layers and photoreceptors

- Small serous elevation of neuroretina
- Small serous elevation of retinal pigment epithelium
- Drüsen
- Polypoids.

TABLE 17	Lesions of inner layers

Lesions of inner layers are observed in case of:
- Arterial occlusions
- Ischemic retinopathy
- Venous occlusions
- Glaucoma
- Neurodegenerative disorders Alzheimer disease
- Neurodegenerative disorders Parkinson disease.

Outline–Segmentation

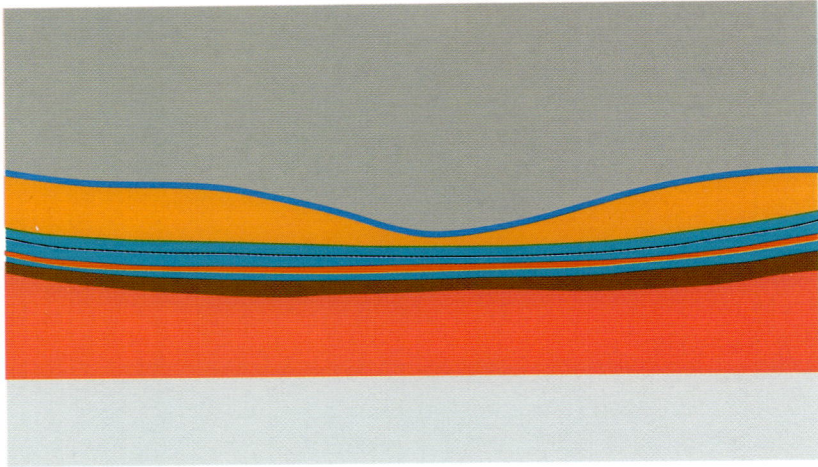

1. Normal retina. The inner retina is represented in orange; the outer retina with external limiting membrane and the inner/outer photoreceptor segment junction are shown in light blue. Below the choroid is visible and then the sclera.

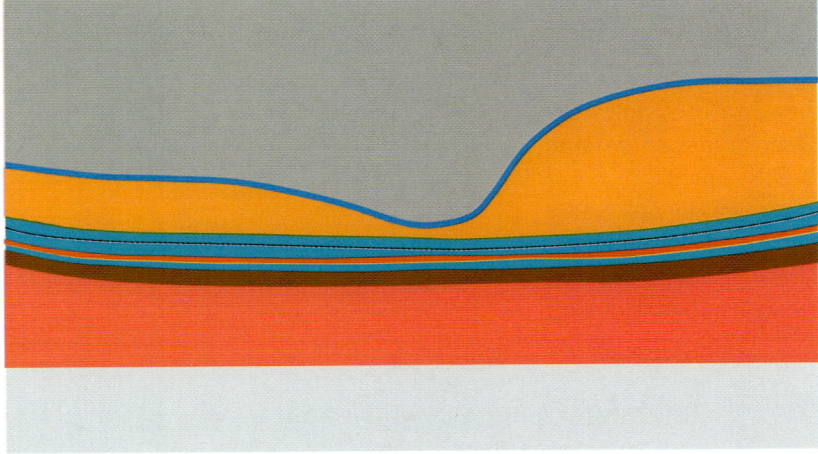

2. Occlusion of an arterial branch with edema of a part of the inner retina.

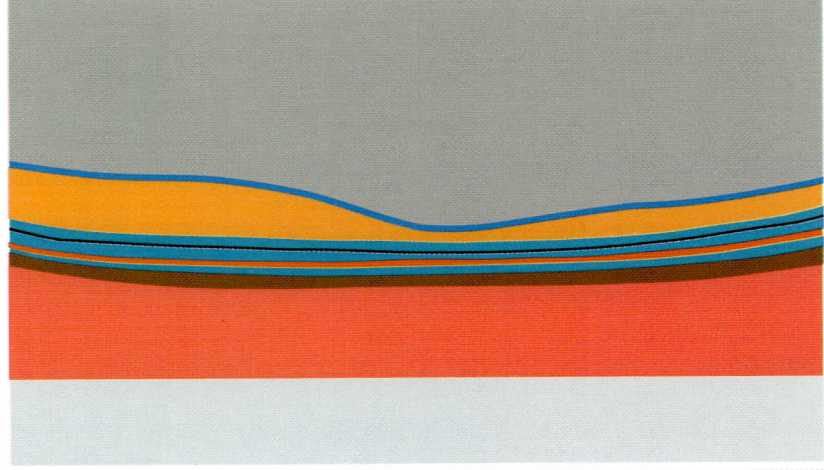

3. Old occlusion of an arterial branch with atrophy of a part of the inner retina.

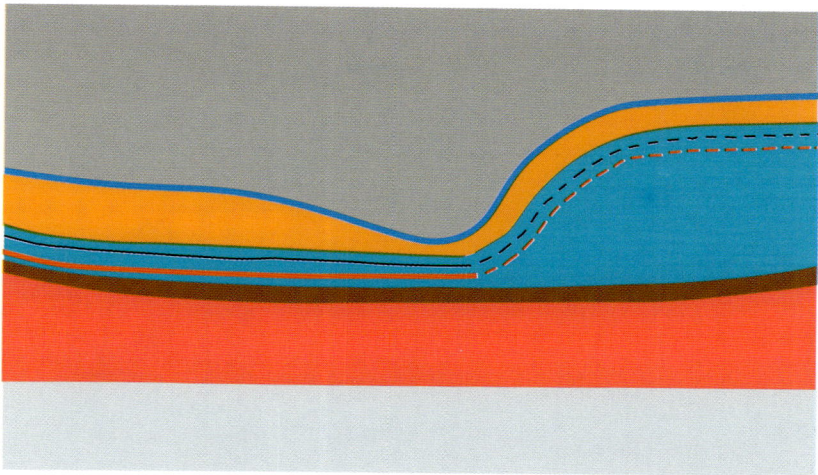

4. Occlusion of a venous branch with edema, localized in a part of the outer retina.

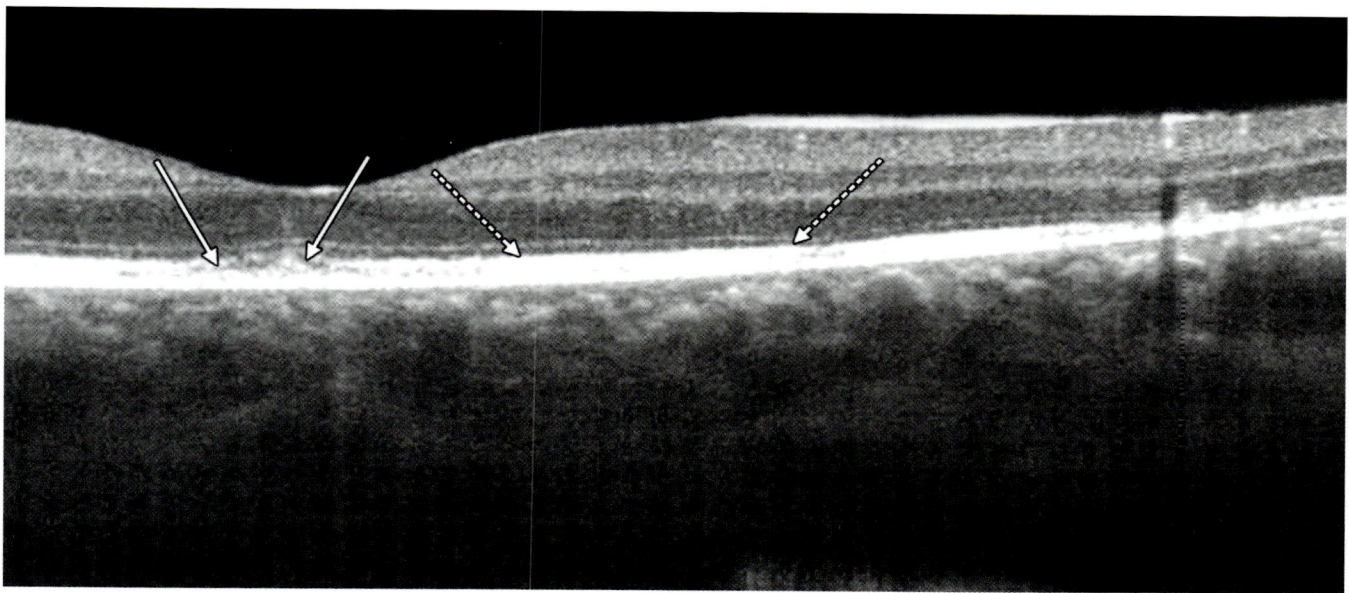

FIGURE 11: ISOLATED LESIONS OF THE ELLIPSOID (INNER/OUTER PHOTORECEPTOR SEGMENT JUNCTION) – MULTIPLE EVANESCENT WHITE DOTS SYNDROME (MEWDS) – MORPHOLOGIC ALTERATIONS SEGMENTATION

An 18-year-old boy exhibited an abrupt loss of 5/10 (20/40) of his vision. The ellipsoid was much altered, interrupted or absent at the fovea level. A slight thickening was present at the level of the external limiting membrane and the outer nuclear layer.

The outer nuclear layer containing the photoreceptor nuclei and the external limiting membrane were normal. The retinal pigment epithelium/choriocapillaris complex was preserved. The choroid presented an increase in vessels diameter. After a month the youth regained a 10/10 (20/20) vision and his OCT alterations disappeared. Line arrows show the lesions of the ellipsoid-junction of the inner/outer photoreceptors segments, dashed arrow shows a slight thickening of the junction.

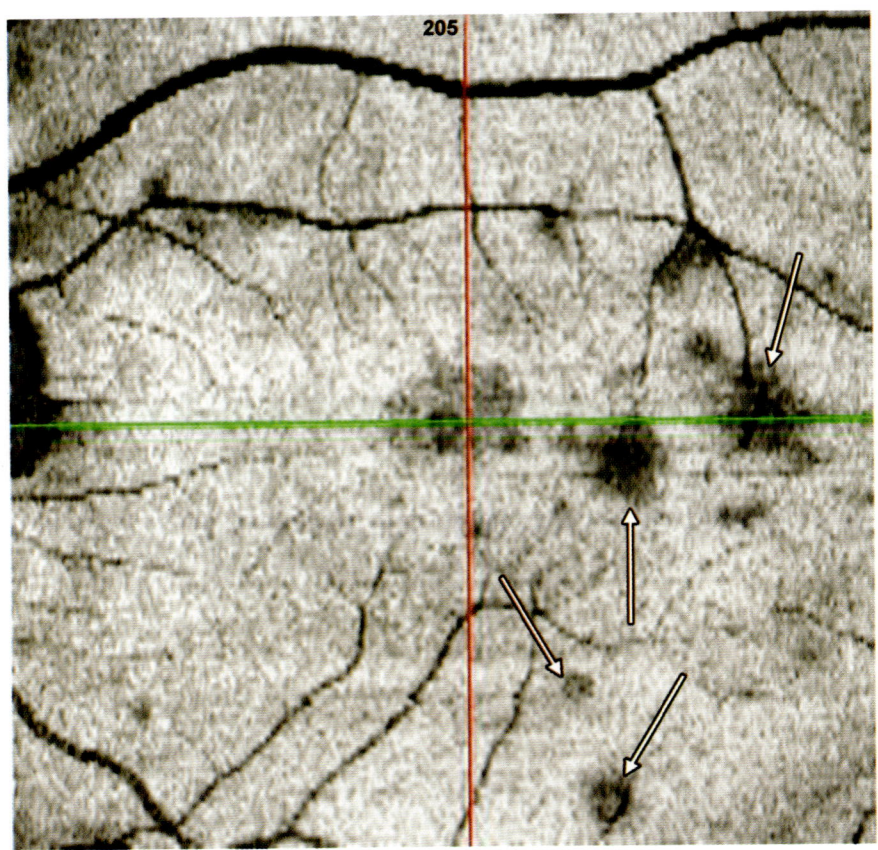

FIGURE 12: LESIONS OF THE ELLIPSOID (INNER/OUTER PHOTORECEPTOR SEGMENT JUNCTION) – "EN FACE" SECTION
The "en face" section had been performed exactly at the level of ellipsoid, using a section thickness of 12 microns, fit to the concave shape of retinal pigment epithelium. The section shows the topography of the matter loss at the junction in correspondence to the fovea, while highlighting the presence of other losses of matter of the ellipsoid, temporally in relation to the macula. Line arrows show the lesions of the ellipsoid-junction of the inner/outer photoreceptors segments.

Reflectivity Study

Layer's reflectivity is one of the fundamental elements of OCT analysis. The reflectivity can be increased, decreased or it is possible to note some shadow areas.

HIGH REFLECTIVITY

High Reflectivity in Inner and Outer Retina

The high reflectivity formations and deposits are discussed in the paragraph dedicated to abnormal formations.

Recent laser coagulations are seen as a vertical, slightly hyperreflective band, from the nerve fiber layer to the retinal pigment epithelium, with photoreceptor alteration.

The retinal pigment epithelium can have a thickness sometimes irregular with the disappearance of one or two layers. The drüsen are indicated by small elevations with a visible Bruch's membrane expressed as a thin line. The elevations (detachment of the retinal pigment epithelium) form with the Bruch's membrane an angle greater than 45 degrees.

The three layers of the retinal pigment epithelium can be evident or not, being fused together.

It is possible to observe the presence of abnormal tissue in contact with the retinal pigment epithelium, provoking a layer deformation: fibrovascular tissue, neovascular active membranes, lipofuscin deposits, Tamoxiphen.

Backscattering

The atrophy of retina and retinal pigment epithelium causes increased light backscattering **(Fig. 13)**.

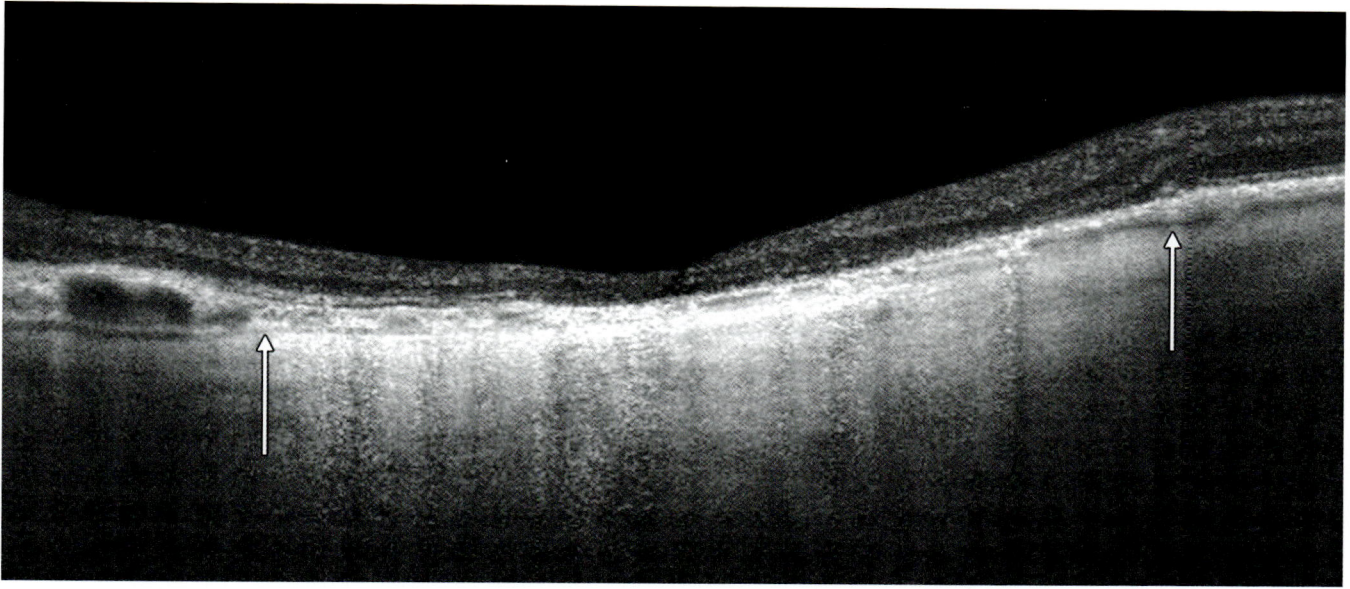

FIGURE 13: RETINAL ATROPHY – DEEP HIGH REFLECTIVITY, DUE TO ATROPHY OF RETINA AND RETINAL PIGMENT EPITHELIUM
The retina is atrophic and very thin. The outer nuclear layer containing photoreceptor nuclei is interrupted at the boundary with the atrophic area, becoming thinner before disappearing altogether. The external limiting membrane and the ellipsoid (inner/outer photoreceptor segment junction) resist for a longer time and are still visible for a few microns inside the atrophic area. The retinal pigment epithelium cells are quite altered and tapered. Bruch's membrane is visible as a thin hyperreflective line. This atrophy of retinal layers and retinal pigment epithelium allows greater light penetration toward the choroid. Sclerotic choroidal vessels are visible and much tapered. Line arrows show the boundaries of atrophic retina.

LOW REFLECTIVITY

Fluid Collection

OCT imaging is especially useful to find, localize and quantify fluid collections, determining their type: serum, blood or other. The fluid collections are shaped and localized in strict relationship with retinal infrastructure and vertical and horizontal retinal barriers.

Fluids are present as diffused edema, cystoid edema, serous or hemorrhagic detachment of neuroretina or retinal pigment epithelium. The collections form cystic or pseudocystic areas, elongated liquid strata in between retinal layers, with sharp limits and variable dimensions **(Tables 18 to 22)**, from minimal to very wide **(Table 23)**.

Intraretinal collection: Edema, cystoid edema, cysts, microcysts, impending hole **(Fig. 14)**.

Subretinal collection: Serous neuroretinal detachment. OCT highlights a retinal elevation at the cone-and-rod tip level and underneath it an optically empty space. The angles formed by the elevation with the retinal pigment epithelium are inferior to 30 degrees. The serous elevation can be idiopathic, associated to acute or chronic *CSCR*, secondary to a serum-hemorrhagic a disciform degeneration or more rarely: myopia, angioid streaks, choroiditis, choroidal neoformations, etc. This collection is often associated with the presence of neovascular membranes within the choroid **(Fig. 15 and Table 24)**.

Pigment epithelium collection detachment: Retinal pigment epithelium detachment. The retinal pigment epithelium is elevationed, separating from Bruch's membrane. The liquid coming from the choriocapillaris accumulates in the subretinal space. Usually, the retinal pigment epithelium detachment forms with the underlying layers an angle of 70-90 degrees, but always greater than 45 degrees **(Figs 16 and 17 and Table 25)**.

TABLE 18	Elevation or cystic space contents
The detachment can be: • Optically empty, in case of serous detachment. Bruch's membrane/choriocapillaris is clearly visible • Hazy, with or without cells, in case of inflammatory exudate or senile degeneration • Hemorrhagic: The elevations highlight optically non-empty contents, much hazy, containing cells with a shadow effect. Bruch's membrane/choriocapillaris is not visible because the blood absorbs the light rays • Filled with fibrovascular tissue with clear fibrous or vascular shoots.	

TABLE 19	Common causes of macular edemas
• Diabetic retinopathy • Venous occlusions (VCR or branches) • Arterial occlusions • Hypertensive retinopathy • Vitreoretinal interface syndrome.	

TABLE 20	Causes of cystoid macular edema

- Diabetic retinopathy
- Venous occlusions
- Chronic retinal epitheliopathy
- Postsurgery edema
- Irvine-Gass syndrome
- Age-related macular degeneration
- Vitreoretinal interface syndrome
- Retinitis pigmentosa
- Uveitis, iridociclitis, choroiditis
- Birdshot retinopathy
- Rare causes.

TABLE 21	Differential diagnosis of macular cystoid edema

Differential diagnosis of macular cystoid edema with:
- Macular retinoschisis
- Hole in progress, impending hole
- Macular holes and pseudoholes
- Microcystic degeneration.

TABLE 22	Diabetic maculopathies

- Maculopathy with diffuse edema
- Maculopathy with cystoid edema
- Serous retinal elevation
- Ischemic maculopathy
- Traction maculopathy
- Associated vascular and traction maculopathy.

TABLE 23	Low reflectivity structures

- Optically empty spaces
- Cysts, microcysts, cavities
- Intraretinal edema
- Exudative neuroepithelium detachments
- Cystoid edema
- Macular telangectasis
- Outer retina tubulations
- Retinal pigment epithelium detachments
- Retinal pigment epithelium hypopigmentation.

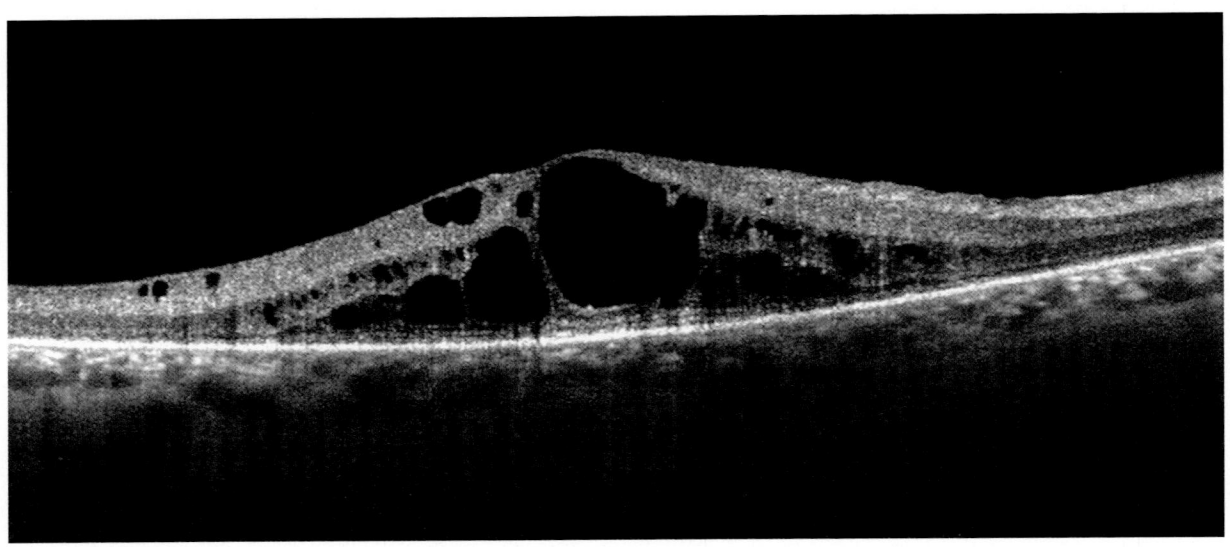

FIGURE 14: MACULAR CYSTOID EDEMA – STRUCTURAL ALTERATIONS – LOW REFLECTIVITY
  The retina is significantly thicker and "dome" shaped, with fluid deposits in the intraretinal small cavities at level of the fovea and Henle's layer. These small cavities appear first in the inner and outer nuclear layers and later, merge, become coalescent and occupy the whole retinal thickness. A larger central cell can be observed, occupying the whole retinal thickness, with preserved but thin retinal surface. The pseudocystic cavity shape is determined by the cavity position in between the retinal horizontal (retinal plexiform parallel layers) and vertical structures.
  The outer nuclear layer, containing the photoreceptor nuclei, is much altered by pseudocysts. The external limiting membrane and the ellipsoid (inner/outer photoreceptor segment junction) are partially altered, but the retinal pigment epithelium/choriocapillaris complex appears normal.

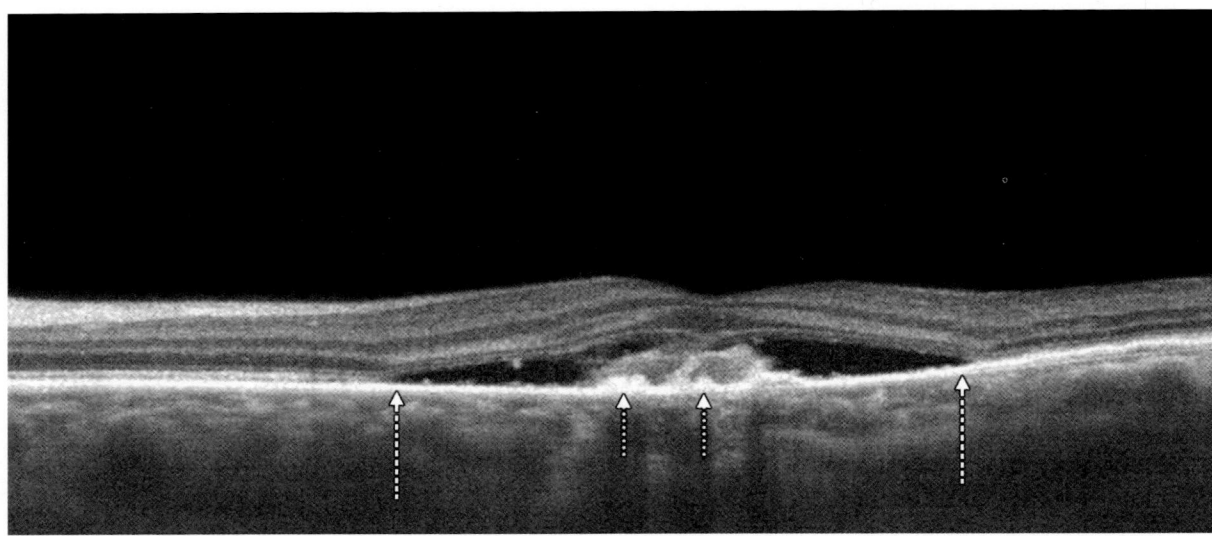

FIGURE 15: NEUROEPITHELIUM DETACHMENT IN A CASE OF CHRONIC EPITHELIOPATHY – STRUCTURAL ALTERATIONS – LOW REFLECTIVITY – OPTICALLY EMPTY SPACES
Below the neuroretinal elevation, we can observe an optically empty space containing deposits of hyperreflective substance, probably lipofuscin. The elevated retina forms an angle of about 15 degrees with the retinal pigment epithelium. The outer nuclear layer, containing the photoreceptor nuclei, the external limiting membrane and the ellipsoid are elevated and the photoreceptor external segments appear slightly altered. Generally, the neuroepithelium elevations form an angle of about 10–30 degrees with the retinal pigment epithelium. Dashed arrows show the limits of the neuroepithelium detachment, the dotted arrows indicate the hyperreflective deposits.

TABLE 24	Causes of serous neuroretinal elevation

- Central serous chorioretinopathy (CSCR)
- Diffuse retinal epitheliopathy
- Subretinal neovascularization
- Polypoidal vasculopathy
- Optic disk colobomatous pits
- Diabetic retinopathy
- Harada disease
- Choroiditis
- Coats disease
- Best's disease, pseudovitelliform degeneration
- Retinal angiomatosis
- Choroidal neoformations (nevi, angiomi, melanomata, metastases)
- Subretinal parasitic cysts

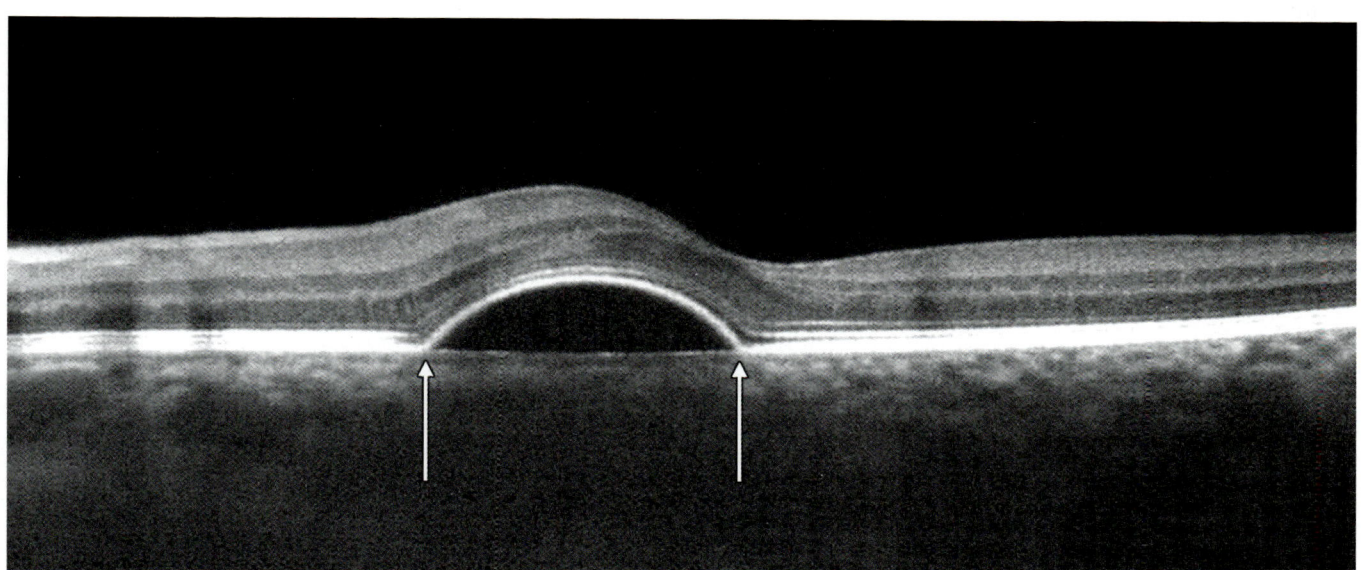

FIGURE 16: RETINAL PIGMENT EPITHELIUM DETACHMENT IN A CASE OF ACUTE EPITHELIOPATHY – STRUCTURAL ALTERATIONS – LOW REFLECTIVITY – OPTICALLY EMPTY SPACES

The fluid coming from the choriocapillaris accumulates between Bruch's membrane and Retinal pigment epithelium forming a domed detachment. The retinal pigment epithelium detachment forms an angle greater than 45 degrees with the Bruch's membrane. The outer nuclear layer, containing the photoreceptor nuclei, the external limiting membrane, the ellipsoid (inner/outer photoreceptor segment junction) and retinal pigment epithelium are preserved and remain parallel to the elevation. The Bruch's membrane appears as a thin hyperreflective line in correspondence to the retinal pigment epithelium elevation. Line arrows show the limits of the pigment epithelium detachment.

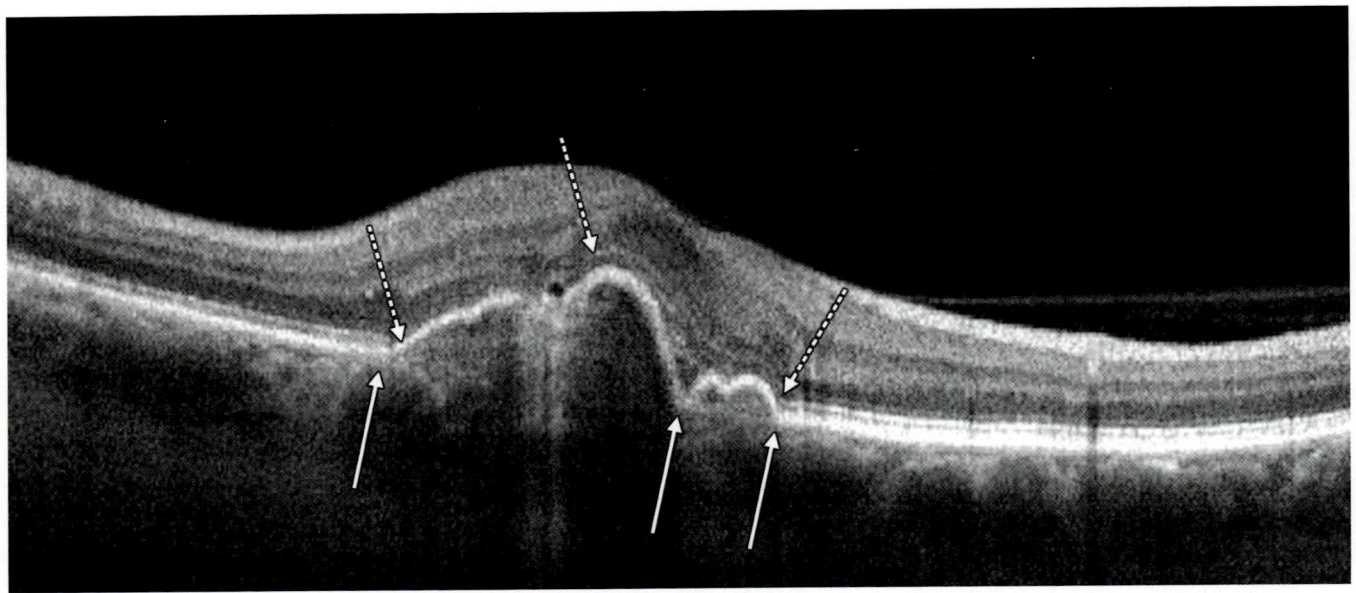

FIGURE 17: RETINAL PIGMENT EPITHELIUM DETACHMENT IN A CASE OF OCCULT NEOVASCULAR MEMBRANE IN AGE–RELATED MACULAR DEGENERATION – STRUCTURAL ALTERATIONS – LOW REFLECTIVITY – OPTICALLY EMPTY SPACES

The retinal pigment epithelium is elevated and separated from the Bruch's membrane, that is not visible. Below the retinal pigment epithelium, slightly hyperreflective, hazy fluid accumulates, determining a very irregular dome-shaped detachment. The outer nuclear layer, containing the photoreceptor nuclei shows a slightly increased thickness. The external limiting membrane and the ellipsoid (inner/outer photoreceptor segment junction) are altered or absent at the elevation level. The neuroepithelium presents a small localized elevation. The occult membrane is not visible on this scan, but is indirectly shown by: slight retinal edema, small neuroepithelium detachment, lesions of external limiting membrane and ellipsoid. Line arrows show the limits of the pigment epithelium detachment, dashed arrows show the limits of photoreceptors junction lesions.

TABLE 25	Causes of serous retinal pigment epithelium detachment
	• Central serous chorioretinopathy (CSCR)
• Diffuse retinal epitheliopathy
• Subretinal neovascularization
• Polypoidal vasculopathy
• Harada disease
• Idiopathic
• Choroidal neoformations (nevi, angiomi, melanomata, metastases)
• Subretinal parasitic cysts
• Miscellaneous causes.

Abnormal Formations

PRERETINAL FORMATIONS

OCT allows assessing density, thickness and location of epiretinal membranes on the retina. **Preretinal or epiretinal membranes** can be more or less thin, due to the presence of fibroglial elements. In the latter case, they are sharply hyperreflective.

The membrane itself can be adherent, fused with the internal limiting membrane, or partially or totally detached. It is more reflective than the normal retinal nerve fibers layer.

The membranes can exert traction on the retina profile, altering it.

Once detached, the posterior hyaloid appears on OCTs as an image slightly reflective, often interrupted, very thin albeit well defined. Sometimes, it adheres to some points of the retinal surface.

The epiretinal membranes can be separated from, in contact or fused with the retina, creating waves or folds on the retinal surface.

Macular pucker: The inner limiting membrane retraction is caused by a fibroglial proliferation, exerting a tangential traction on its interior. Formed by glial tissue adhering to the vascular arches, a large preretinal membrane has contracted, creating retinal folds. Generally, the folds are transversal to the retina below the retracted membrane, and radial outside the retraction area.

Cotton wool exudates: These retinal formations have a hyperreflective surface, and are in contact and fused with superficial retinal layers such as the nerve fiber layer. They have variable size and are linked to nerve fiber lesions, at the periphery of ischemic areas.

INTRARETINAL FORMATIONS

Hemorrhages: When dense, they create shadows.

Hard exudates: Localized in the posterior layers of the retina, they are formed by lipoproteins and can be isolated or part of circinate retinitis. Generally, they precipitate in the outer plexiform and nuclear layers, at the divide between edematous and normal retina. Their shape and positioning are determined by retinal structure. In more advanced cases, these exudates can also precipitate in the inner plexiform and nuclear layers **(Fig. 18 and Tables 26 to 28)**.

POSTERIOR FORMATIONS

When a macular degeneration or a diabetic fibrous vascular plaque reaches its terminal evolution, fibrous scars appear as a markedly hyperreflective nodule, deforming the retinal profile and eliminating all normal elements of the retina. Scars often present pigment deposits, projecting dense shadows onto the posterior layers.

Generally, the neovascular membranes in young and myopic subjects and the classical membranes associated to recent age-related macular degenerations are fusiform, nodular, rounded. They are located immediately in front of the retinal pigment epithelium layer. When active, they are always linked to edema or serous neuroretinal detachments. When evolved for some months, the neovascular

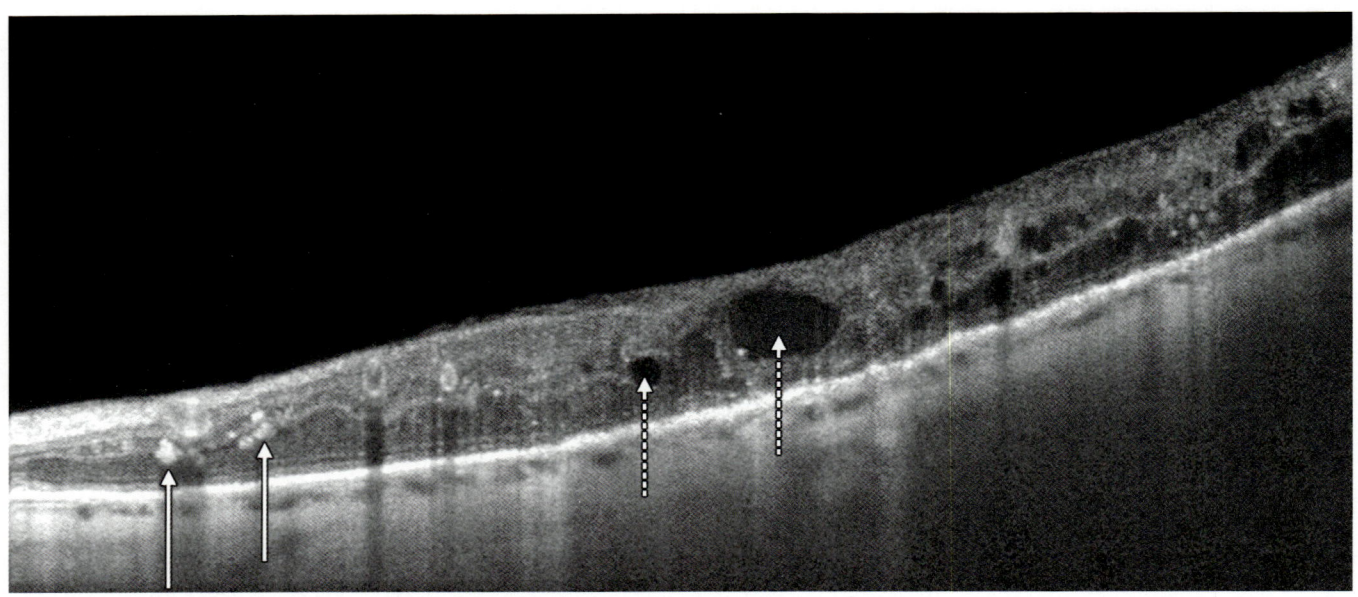

FIGURE 18: EXUDATES – ABNORMAL FORMATION
Hard exudates can be isolated, as in this case of diabetic retinopathy, or be part of a circinate retinitis. We can observe deep hyperreflective formations, composed of lipoproteins, which are deposited at the divide between edematous and normal retina. There, they are located in the inner and outer nuclear and plexiform layers, although they can infiltrate and obstruct every other layer. The exudates shape and location are determined by their situation between parallel horizontal retinal barriers and vertical retinal structures. The external limiting membrane, the ellipsoid (inner/outer photoreceptor segment junction) and the retinal pigment epithelium/choriocapillaris complex are interrupted by a shadow (screen effect) projected onto the posterior layers. We can further observe, in some points, an actual alteration of both external limiting membrane and ellipsoid. One should also note a thin, very adherent epiretinal membrane. The choroid is very thin. Line arrows show hard exudates, dashed arrows show cystoid edema cavities.

TABLE 26	Abnormal formations
	• Preretinal
	• Epiretinal membranes
	• Intraretinal
	• Deep formations.

TABLE 27	Causes of cotton wool exudates
	• Diabetic retinopathy
	• Hypertensive retinopathy
	• Gravidic toxemia
	• Anemia
	• Leukemia, Hodgkin's disease
	• Purtscher retinopathy.
	Rare causes
	• Kala-azar
	• Leptospirosis
	• Disseminated lupus erythematosus.

TABLE 28	Causes of hard exudates
• Diabetic retinopathy • Hypertensive retinopathy • Radiation retinopathy • Coats' disease • Exudative macular degeneration.	

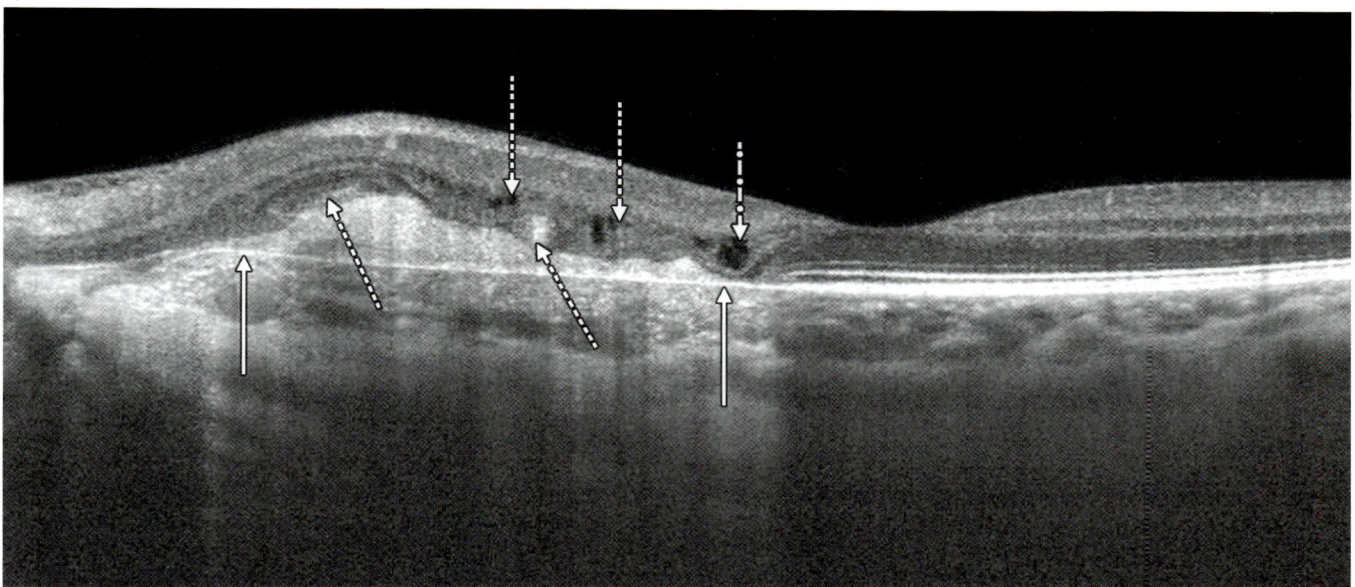

FIGURE 19: CLASSICAL NEOVASCULAR MEMBRANE – DMLE – ABNORMAL FORMATION
An extensive formation with irregular reflectivity is localized in front of the altered retinal pigment epithelium and Bruch's membrane. In front of this active neovascular membrane we can see a macular edema with some small cystoid chambers. The retinal thickness is increased by a diffuse edema and some optically empty cavities of cystoid edema at the nuclear layers level. The outer nuclear layer containing the photoreceptor nuclei, the external limiting membrane and the ellipsoid (inner/outer photoreceptor segment junction) are destroyed and the retinal pigment epithelium is thin and disaggregated. The rounded cavity at the scan center seems to be an outer retina tubulation. The choroid appears slightly sclerotic but its thickness is almost normal. Line arrows show the limits of the neovascular membrane, dashed arrows show the cystoid edema cavities, dotted arrows show the hard exudates, dotted-dashed arrow indicates outer retinal tubulation.

membranes are more difficult to identify, appearing as a thickening of the retinal pigment epithelium/ choriocapillaris complex, with a disaggregated look, and they are associated to edema or sercus detachments of the overlying retina **(Figs 19 and 20)**.

The occult neovascular membranes are difficult to identify: they may appear as an irregular thickening of the retinal pigment epithelium/choriocapillaris complex with a certain degree of retinal pigment epithelium disaggregation. When active, they are always associated to edema or serous neuroretinal detachments **(Tables 29 to 31)**.

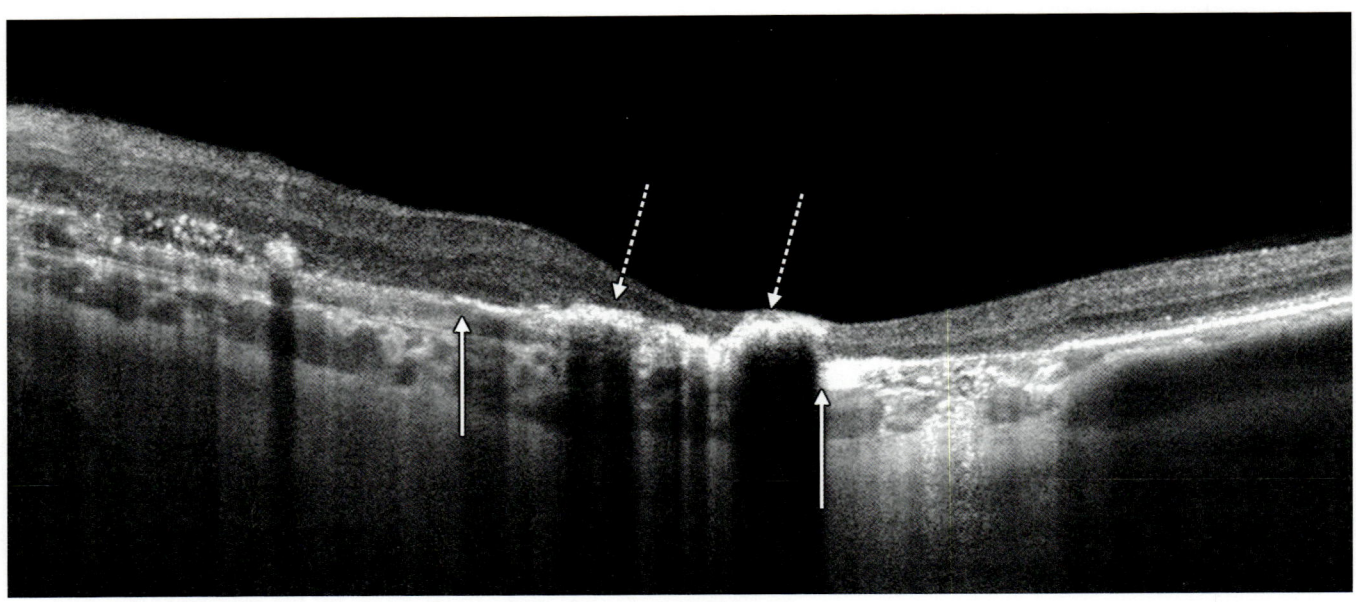

FIGURE 20: PIGMENT DEPOSITS – ABNORMAL FORMATION
Chorioretinal atrophy area caused by an old scar. There is a marked penetration of light rays towards the posterior layers, at the atrophic area level. Dense pigment deposits are visible and form a screen, thus determining a shadow onto the sclera. Line arrows show the limits of the dense scar, dashed arrows show the pigment deposits.

TABLE 29	Causes of neovascular membranes
• Age-related macular degeneration • Chronic retinal epitheliopathy • Myopia • Uveitis, iridocyclitis, choroiditis • Birdshot retinopathy • Pseudovitelliform degeneration • Nevi • Osteomata • Rare causes.	

TABLE 30	Abnormal structures and formations
High reflectivity • Substance deposits (lipofuscin, tamoxiphen) • Retinal pigment epithelium hypertrophy • Nevi • Scar tissue • Hemorrhages • Neovessels • Hard exudates.	Neovascular membranes • Fusiform • Elongated • Rounded • Nodular. **Medium reflectivity** • Retinal edema. **Low reflectivity** • Cavities, cysts • Detachments, elevations, schisis • Projected shadow.

TABLE 31	Analytic qualitative study of high reflectivity abnormal formations, by retinal level

Surface formations
- Epiretinal or intraretinal fibrosis
- Cotton wool exudates.

Intraretinal formations
- Inflammatory infiltrates
- Hemorrhages
- Hard exudates.

Deep formations
- Lipofuscin deposits
- Tamoxiphen deposits
- Pigment epithelial hyperplasia
- Retinal pigment epithelium atrophy with backscattering
- Pre-epithelial and subepithelial neovascular membranes.

Shadow Areas (Screen Effect, Shadow Cone)

A dense tissue can act as a total or imperfect screen, creating a shadow area that hides the posterior elements.

The retinal structures can be concealed at the preretinal, intraretinal or posterior levels. The shielding elements are generally single, but can be associated **(Table 32)**.

Normal shadow cone: Normal retinal vessels can be seen as rounded or *hourglass-shaped* formations, creating a shadow cone that interrupts the inner/outer photoreceptor segment junction.

Anterior retina: Very dense, thick, superficial and deep hemorrhages cause a shadow cone. The same effect is also achieved by cotton wool exudates.

Deep retina level: Hard lipidic exudates within the deep retinal layers project a dense shadow.

Retinal pigment epithelium: Thickening, hyperplasias and hypertrophies of the retinal pigment epithelium: the pigment formations, hyperplasias and hypertrophies of the retinal pigment epithelium are very dense, as much as certain nevi and melanomas.

Scar: A dense scar with pigment deposits generates an intense shadow area.

When they are thick, highly hyperreflective subretinal neovascular membranes form a shadow area.

Choroid: The choroidal nevus is expressed by a reflectivity increase below the retinal pigment epithelium-Bruch's membrane complex, projecting a shadow onto the posterior layers **(Fig. 21)**.

TABLE 32	Shadow effect—screen effect

Anterior retina
- (Normal) retinal vessels
- Hemorrhages
- Exudates
- Laser spot.

Posterior retina
- Retinal scars
- Neovascular membranes
- Hyperplasias and hypertrophies of the retinal pigment epithelium
- Thickening of the retinal pigment epithelium
- Pigment deposits
- Choroidal nevus.

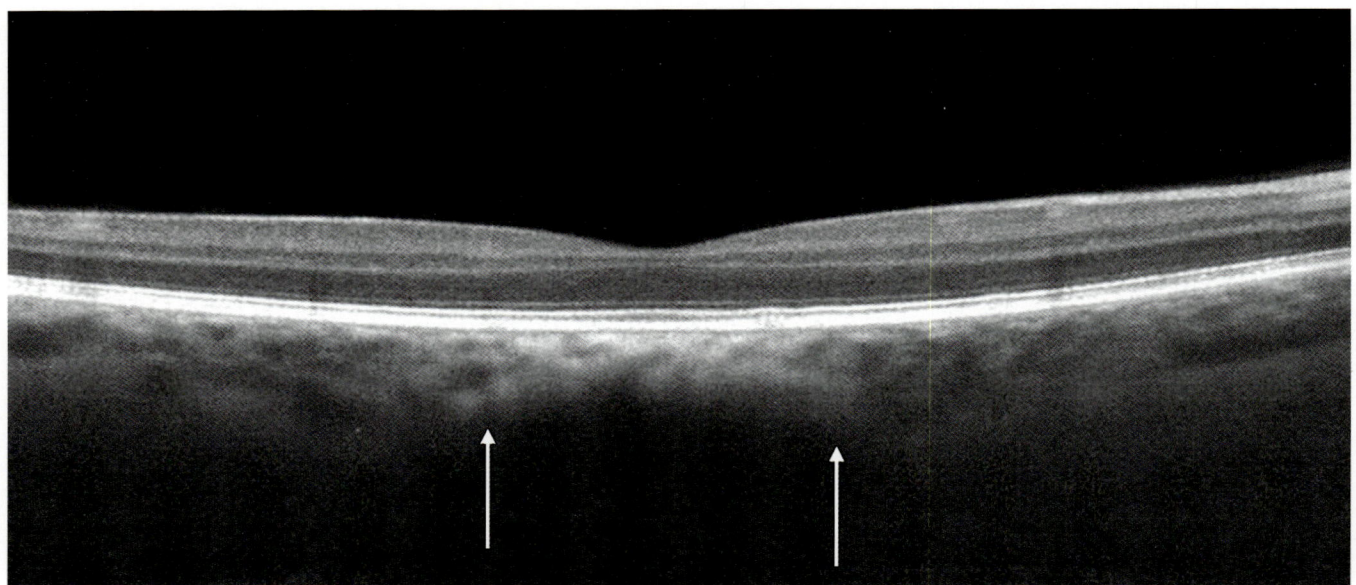

FIGURE 21: CHOROIDAL NEVUS – DEEP HIGH REFLECTIVITY AND PROJECTED SHADOW
In this case of choroidal nevus, at the fovea level, we can see a normal outer retina, with normal photoreceptor layer, internal limiting membrane and ellipsoid (inner/outer photoreceptor segment junction). The retinal pigment epithelium/choriocapillaris complex is normal as well. Below the retinal pigment epithelium, the **choroid presents a more reflective area, i.e. the nevus**. Its hyperreflectivity is caused by a pigment deposits, generating a shadow underneath that is projected onto the posterior choroidal layer. Line arrows show the limits of the choroidal nevus.

CHAPTER
4
Quantitative Analysis of Pathologic OCTs

The purpose of quantification is to compare numeric data on a given disease's evolution, both spontaneous and after medical, laser or surgical therapy.

Linear Measurements of Axial B-scans

Most software programs trace at least two markers on the retinal tissue. The first marker corresponds to the vitreoretinal interface and is common to all software programs, because it divides two very high contrast structures (the vitreal "silence" and the nerve fiber layer). The second marker should delimit the retinal pigment epithelium. As it is well known in OCT Spectral Domain, the retinal pigment epithelium is constituted of several layers and is in fact defined as the retinal pigment epithelium/choriocapillaris complex. Therewithin we can observe the interdigitation between photoreceptor external articles and retinal pigment epithelium villosities, cell layer of retinal pigment epithelium and Bruch's membrane. That complex measures up to 60 microns and is constituted by high and low reflectivity layers. The software can trace the marker to its external or internal limit, or within the complex. Therefore, the measure of retinal thickness can vary depending on the software, to the point of inhibiting data comparisons.

A further marker is traced at the inner plexiform layer level and separates the neuroepithelium in outer and inner retina.

Some software programs of the latest generation trace further markers at the retinal pigment epithelium/choriocapillaris complex level.

Measurement of Coronal Images and "En Face" Scans

The acquisition of a series of linear B-scans, parallel top to bottom to the macula area, allows reconstructing the macular cube and dissecting it by slices that are planar or fit to the curvature.

In terms of quantitative analysis of macular thickness and volume, the software uses the calculated surface, interpolating the markers of each B-scan that forms the macular cube.

Thus we can calculate the maps of outer retina, internal and full thickness. Exploiting the additional markers within the retinal pigment epithelium/choriocapillaris complex, we can study the drüsen morphology and the appearance of the retinal pigment epithelium alterations, subtracting the whole overlying neuroretina.

Some instruments extend 3D calculation capabilities even at the deeper layers, outside the retinal pigment epithelium screen. Thanks to the particularly large bandwidth of the spectrometer, in the 89% of cases, the Optovue RTVue100, Optovue XR and Heidelberg software can visualize the choroidal layers up to the suprachoroidal space (lamina fusca). The small vessels (Sattler) layer and the large vessels (Haller) layer can be highlighted using coronal sections fit to the average curvature of the posterior pole. Sattler's layer can be seen reducing the thickness of the slice (<15 microns) and locating it just below the retinal pigment epithelium/choriocapillaris complex (0–40 microns), whereas Haller's larger vessels become visible by increasing the slice thickness (>20 microns) and scanning at a greater depth (100–120 microns below the retinal pigment epithelium/choriocapillaris complex). An even deeper scan sometimes yields images of the arteries (characterized by a distinctive "hook-shape").

At this time, software is being developed to allow setting calipers along the three axes.

Quantitative Segmentation

The segmentation of the nerve fiber layer was the first to be automatized by all of the OCT instruments for the purpose of studying glaucoma and its evolution.

Fast and easy to use, the OCT device generates data on the inner and the outer retina, i.e. inner retina, ganglion cell complex; outer retina including all layers between retinal pigment epithelium and inner plexiform layer. The segmentation data are not reliable in the presence of significant retinal alterations.

Thanks to segmentation, we can measure the inner and outer retina thickness and volume.

We also study the localized alterations of the retinal layers. These layers need to be measured one by one and each should be compared with the others and subsequently with the contralateral (adelphous) eye.

Retinal Mapping

The Spectral Domain technology has significantly improved the precision and repeatability of macular maps, due to its high acquisition speed. Spectral Domain instruments acquire from 8 to 60-80 B-scans per second, achieving a good resolution. It is possible to get over 140 scans in 4 seconds, thus creating a good retinal map. In fact, we can execute at least 12 radial scans in less than 1 second and 64 horizontal scans, from top to bottom (or "raster scan") in about 1.5 seconds.

The macular maps vary according to the required degree of precision.

The Optovue instrument produces retinal maps imaging a 4 × 4 mm cube of tissue, with 141 horizontal aligned scans. This cube is delimited by the internal limiting membrane on one side, and by the retinal pigment epithelium on the other. The retinal topographic map is subdivided in sectors with indications of the average thickness and standard deviation from a normal sample. Optovue allows subdividing inner and outer retina. In the latter case, the inner plexiform layer is taken as limit. Therefore, we can acquire a full thickness retinal map, or an internal or external retinal map. These maps can be bi- or tridimensional.

The coronal maps can show a grid or the level curves. At the same time, we can visualize on the screen the sagittal, horizontal and vertical scans, and select them with the mouse.

We can develop a map of all altered elevations of the retinal pigment epithelium vs a 3D ovoid representation of the reference retinal pigment epithelium.

These maps can be compared with up to 5 tests, selected by the operator, pertaining to the same patient.

We can also measure:
- Thickness of the nerve fibers layer
- Thickness of the ganglion cells layer
- Profile of each scan
- Average thickness of the nerve fibers
- Map of the nerve fiber layer **(Figs 1A to C)**.

CAUTION! The retinal map can exhibit marked aberrant variations in some sectors, to the point of making its interpretation impossible. These variations can depend from software errors. In case of marked hyporeflectivity or significant alteration of the tissue curvature, the software cannot exactly localize the retinal surface, as it happens in the presence of retinal edema, epiretinal membrane far from the retinal surface or deep staphylomata. In these cases, the software reproduces increases or reductions in thicknesses that do not represent the real situation.

Quantitative Analysis: Thickness

INCREASED THICKNESS

Edema is the major cause of retinal thickening. The study and evaluation of both the various forms of retinal edema and the liquid collections represent two of the greatest advantages offered by OCTs.

REDUCED THICKNESS

It can be seen in the case of atrophic macular degeneration, which produces atrophic areas within a retinal pigment epithelium that is reduced in thickness and reflectivity following the pigment loss.

The hyporeflective retinal pigment epithelium allows a greater light penetration in the choroid. The reflectivity increases due to a reduced light adsorption at the retina and retinal pigment epithelium level.

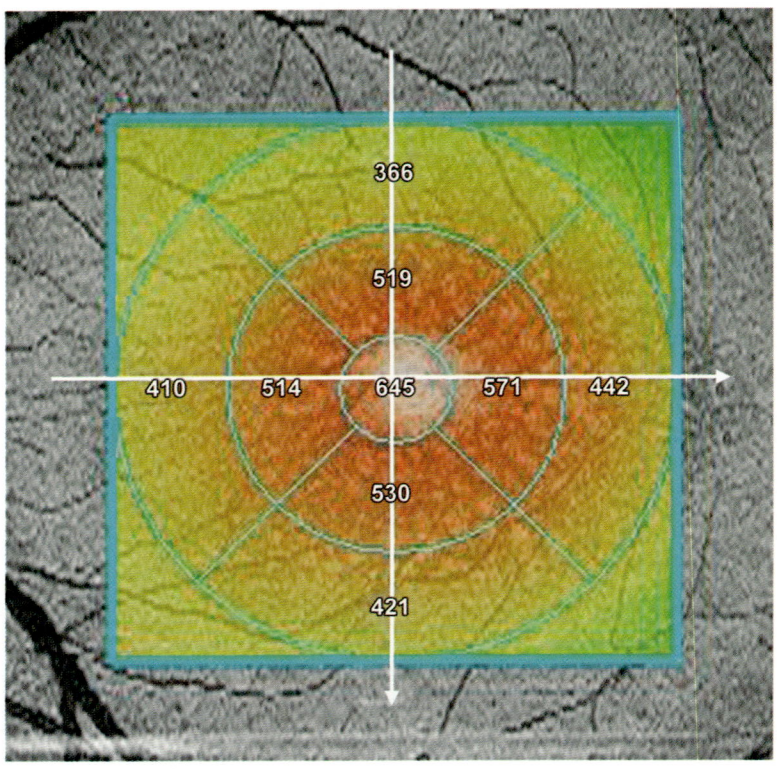

FIGURE 1A

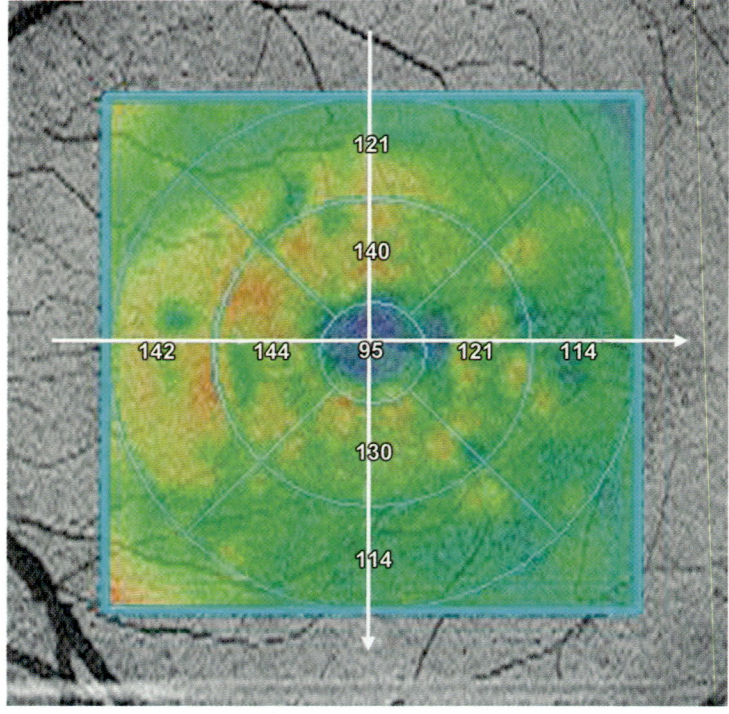

FIGURE 1B

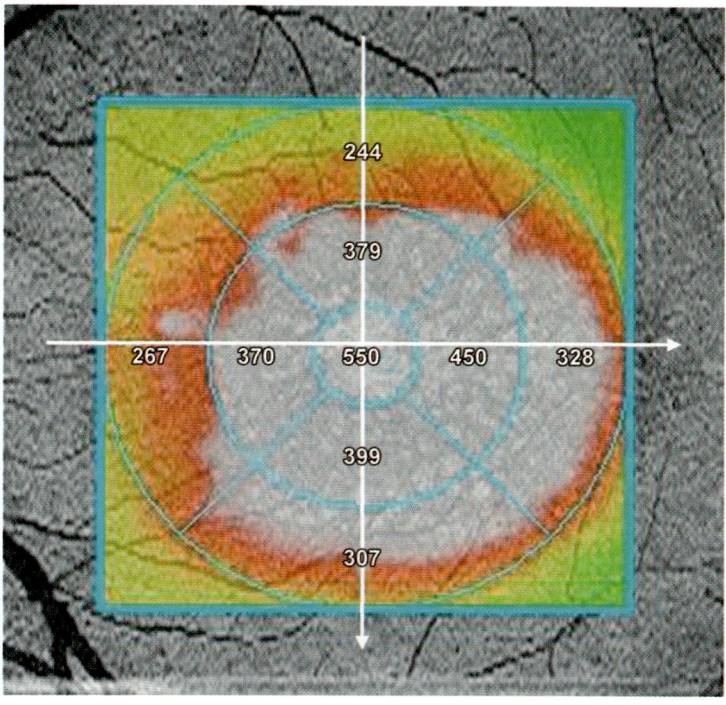

FIGURE 1C

FIGURES 1A TO C: MACULAR MAPPING
Visualizations of a full thickness macular map (A), inner retinal map (B) and outer retinal map (C) in a case of juvenile macular schisis. The outer retinal map shows a marked thickening.

THICKNESS VARIATIONS OF THE RETINAL LAYERS

It is important to assess the thickness of each retinal layer and in particular the outer nuclear layer, containing the cone and rod nuclei. The retinal layer thickness varies with the pathology. Different scans and the retinal map allow pinpointing the layer localization and evolution. In the case of glaucoma, it is essential to study the variations of retinal fiber thickness using chorioretinal circular scans at the papilla level. In the case of glaucoma and neurodegenerative disorders as Alzheimer and Parkinson diseases, it is essential to study the variations of ganglion cell layer thickness.

Retinal Volume

Simple calculations allow defining the volume of the posterior retinal pole or that of some sections.

Knowing the retinal map and its thickness, the instrument software calculates and prints the retinal volume for each map. This is extremely important for the purpose of following the evolution of an edema or a serous detachment, both in the case of spontaneous evolution, and after laser or surgical intervention. The OCT is also indispensable to the study of the pharmacological effects of various medicinal substances.

Optovue and other devices software allows delimiting certain area and calculating their volume **(Tables 1 and 2)**.

Quantitative Choroid Analysis

Currently, we can measure thickness in several positions.

Quantitative Optic Disk Analysis

OCTs allow measuring and quantifying the papillar excavation (depth and surface), the optical disk (surface), the surface of the papillar margin, its volume, the relationship of optical disk/excavation and other elements that will be discusses in another manual.

TABLE 1	Quantitative analytical retinal study
Linear measurements **Retinal thickness** • Increased • Decreased • Variation of one layer thickness (i.e. nervous fibers, ganglion cell layer). **Quantitative segmentation** • Retinal mapping • Volumetry.	

TABLE 2	Choroidal thickness variations
Physiological causes of reduced choroidal thickness • Age • Myopia. **Pathologic causes of reduced choroidal thickness** • Atrophic macular degenerations • High myopia • Glaucoma. **Pathologic causes of increased choroidal thickness** • Venous occlusions • CRSC • ERD.	

CHAPTER
5 "En Face" Imaging

3D images form an important part of the OCT eye study. Current OCTs allow us to acquire three types of images:
- Tridimensional images
- Frontal plane scans. Frontal scans are clinically known as "en face" scans. They are sometimes called coronal or transverse scans.
- Frontal "en face" scans fit (adapted) to the retinal pigment epithelium concavity.

The meaning of data thus obtained is not always evident to the ophthalmologist. In effects, the retinal pathology is seen in a different view and at different angle, which can be disconcerting at first. Albeit the initial interpretation might be difficult, the new aspects soon become easly understood and their clinical usefulness evident.

3D Images: Segmentation of Chorioretinal Layers

OCT device allows acquiring a series of B-scans, forming a tissue cube that can be examined and sectioned from many sides.

OCTs can segmentate the chorioretinal layers. Each layer is separately identified on the basis of its reflectivity. Specific keys allow subtracting some layers and highlighting others, in particular those of nerve fibers, ganglion cells, inner and outer retina, retinal pigment epithelium and choroid. Thus, we can obtain 3D images of the retina or retinal pigment epithelium surface.

The software highlights a sheet representing the internal limiting membrane, a second one representing the retinal pigment epithelium and a third that reconstructs the whole retina, from the internal limiting membrane to the retinal pigment epithelium. Inner retina and outer retina can be separately acquired. The inner plexiform layer marks the limit between inner and outer retina.

These 3D images are very useful for teaching purposes.

Frontal Plane Scans

Certain OCT instruments produce only frontal (coronal), "en face", perfectly plane sections. The series of sections thus created goes from the vitreous to the choroid. Considering that the fundus is cup-shaped, each section cuts different retinal layers.

The images thus acquired are interesting, but can initially be difficult to understand.

Frontal "En Face" Sections Fit to the Retinal Pigment Epithelium Concavity

OCT "en face" images are a clear progress vs the plane sections. In effect, the posterior ocular pole is cup-shaped and each retinal layer is fit to this shape. The software uses the retinal pigment epithelium to calculate the ideal eye curvature. Thus, we obtain an ideal concave surface, the basis to separately highlight the retinal layers.

The transversal or "en face" sections are not plane, but fit to the tridimensional paraboloid concavity of the retinal pigment epithelium.

The frontal, "en face" retinal section exactly follows this concavity and can be moved closer to the sclera or the vitreous, according to the need **(Figs 1 to 8, Table 1)**.

Thickness of the Retinal Section

We can modify the thickness of the retinal section from 1 to 90 microns. Reducing the thickness we increase the sensitivity, but the images become less sharp. When we increase the thickness, the images are sharper, but the sensitivity is reduced and some details are lost. Then, each section can concern a single retinal layer or more layers, according to the thickness. Physicians can vary the section thickness according to the structure to be studied.

The frontal, "en face" images should be studied in series, allowing us reconstructing a third dimension that is particularly useful to study and measure the retinal structures.

3D images are very useful for teaching or explaining the disease to the patient, whereas the "en face" are important for diagnostic and treatment purposes, their interpretation difficulties notwithstanding.

TABLE 1	Clinical interest of 3D OCTs

<ul><li>Epiretinal membranes, macular pucker</li><li>Macular profile</li><li>Diffuse retinal edema</li><li>Irvine-Gass retinal cystoid edema, diabetes</li><li>Macular telangiectasis</li><li>Macular holes</li><li>Lamellar holes</li><li>Retinoschisis</li><li>Vitreomacular traction—retinal folds</li><li>Drüsen</li><li>Neovascular membranes</li><li>Retinal pigment epithelium detachment</li><li>Choroid study
  - Choriocapillaris study
  - Sattler's layer study
  - Haller's layer study</li><li>Sclera study: Intrascleral channels, short ciliary arteries, Tenon's capsule.</li></ul>	

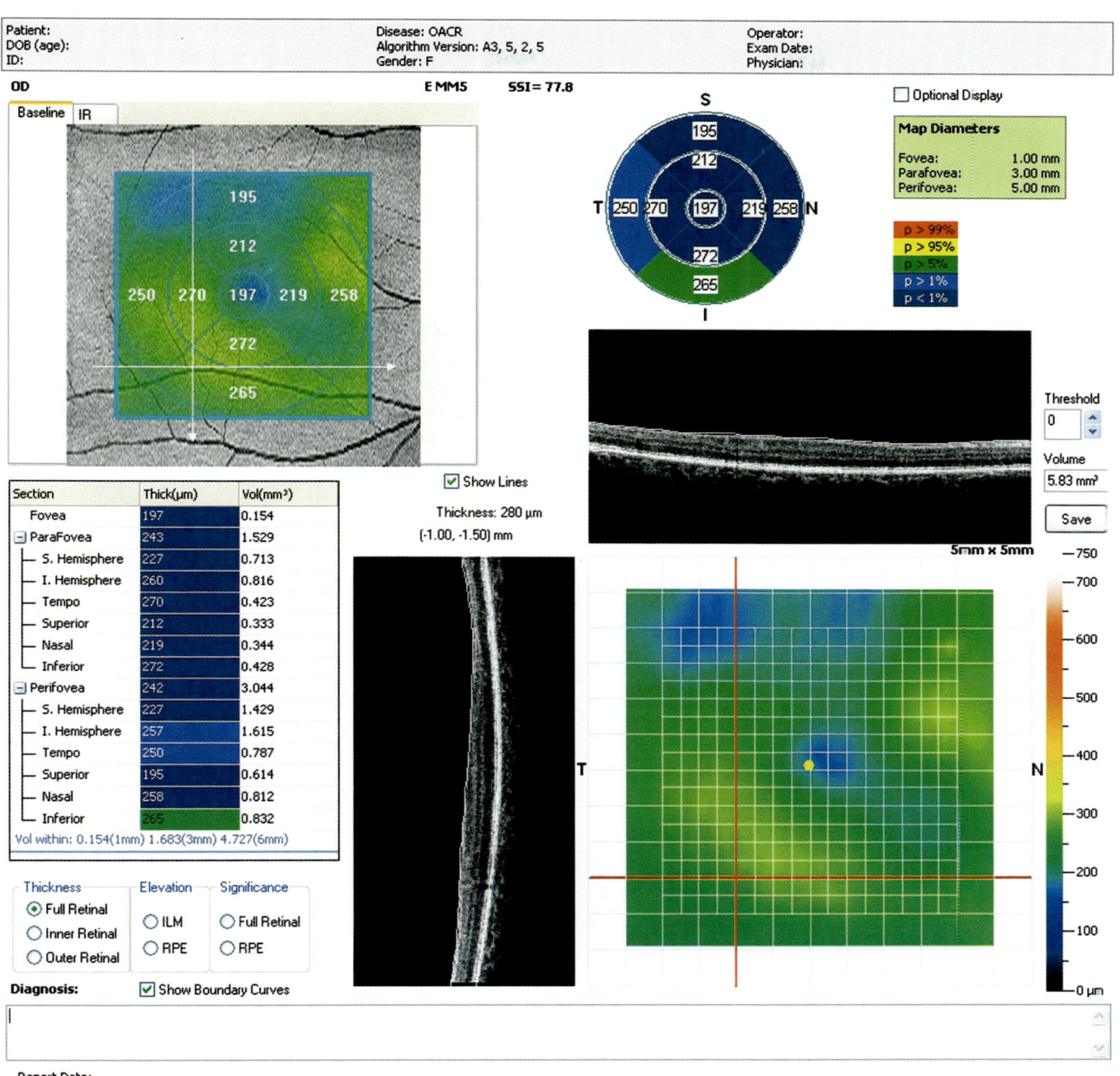

FIGURE 1A

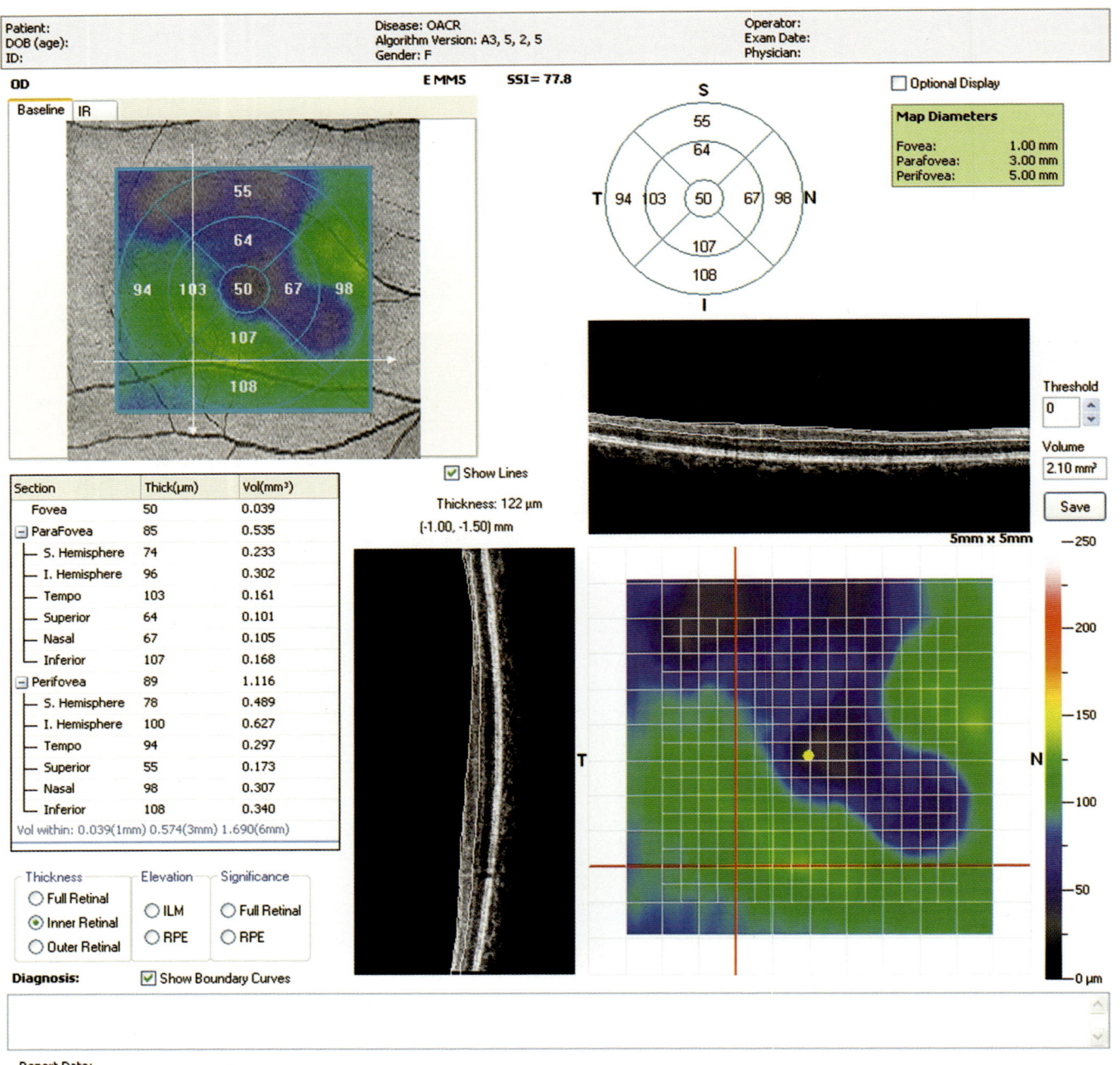

FIGURE 1B

"EN FACE" IMAGING

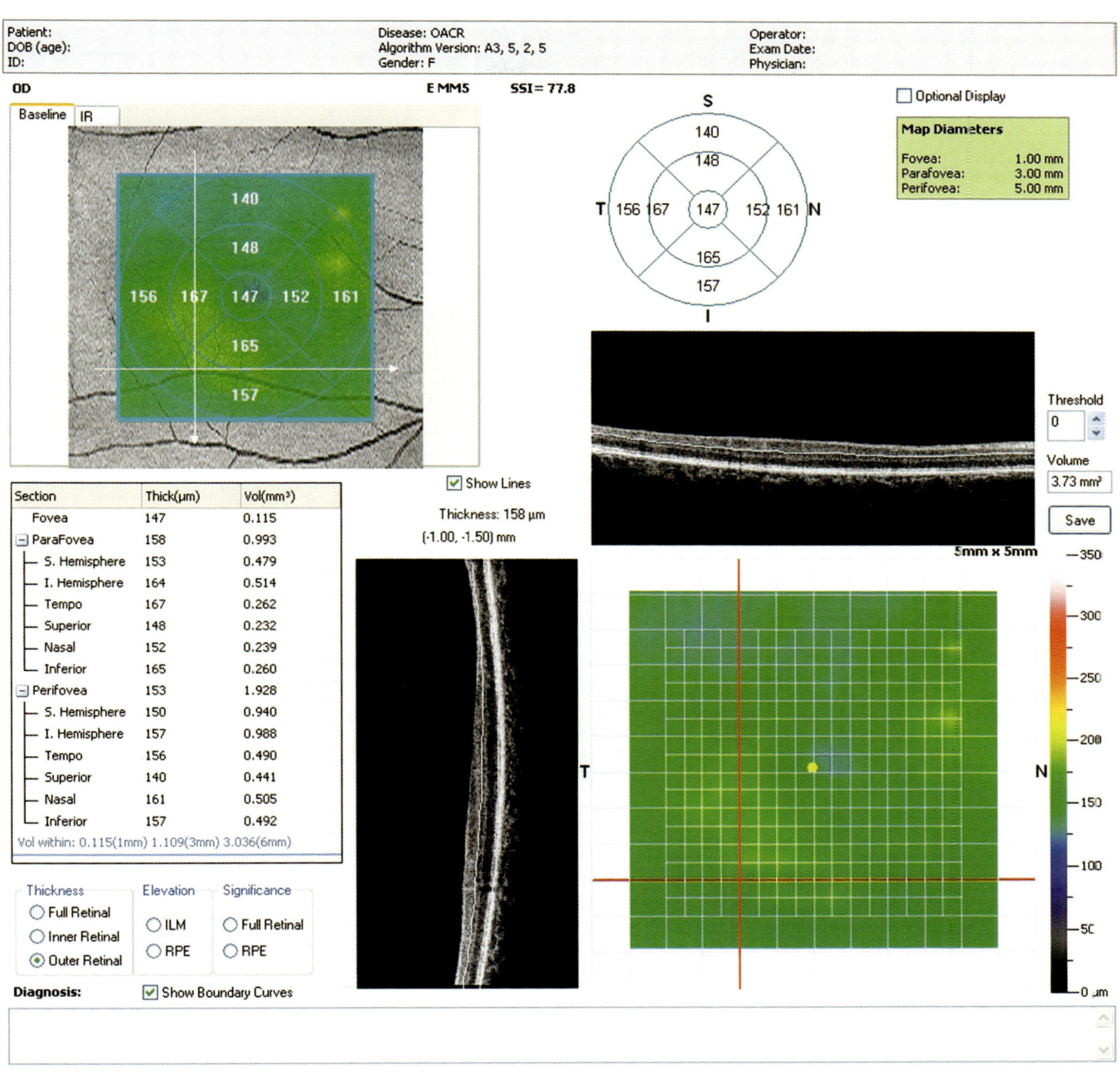

FIGURE 1C

FIGURES 1A TO C: RETINAL TOPOGRAPHY – OCCLUSION OF AN ARTERIAL BRANCH
The retinal maps are acquired with Optovue Rtvue, starting from a 4 × 4 mm tissue cube. This cube is referred to the internal limiting membrane and the retinal pigment epithelium. The inner plexiform layer marks the divide between inner and outer retina. The retinal topographic map is divided in sectors with indications of the average thickness. In this case, we see three maps: a full thickness retinal map (A), the inner retina thickness (B) and the outer retina thickness (C). In this case of arterial branch occlusion, it is interesting to note the occlusion level.
A: The retina, completely visible, is slightly altered.
B: The inner retina presents a marked atrophy.
C: The outer retina is normal.

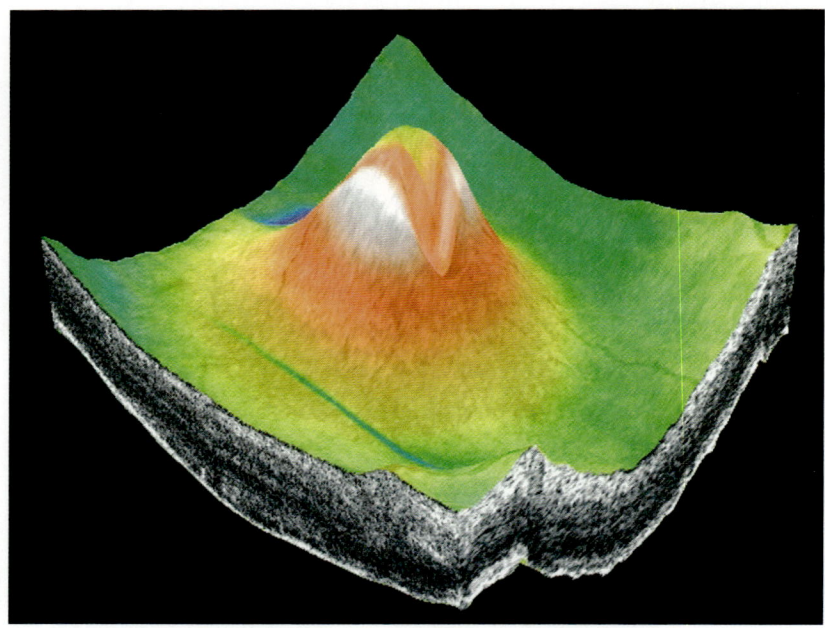

FIGURE 2: 3D IMAGE
3D view of a retinal pigment epithelium detachment, using the topographic software of the instrument. 150 B-scans reconstruct a chorioretinal 3 × 3 mm cube, which can be observed from different directions. A command allows eliminating the various layers, starting with the vitreous and then erasing in sequence nerve fibers and retina. Thus, we can study in 3D the surface of the retinal pigment epithelium, highlighting a regular detachment in a case of central serous chorioretinopathy. This image is very useful for teaching purposes and to explain the patient his pathology.

"EN FACE" IMAGING

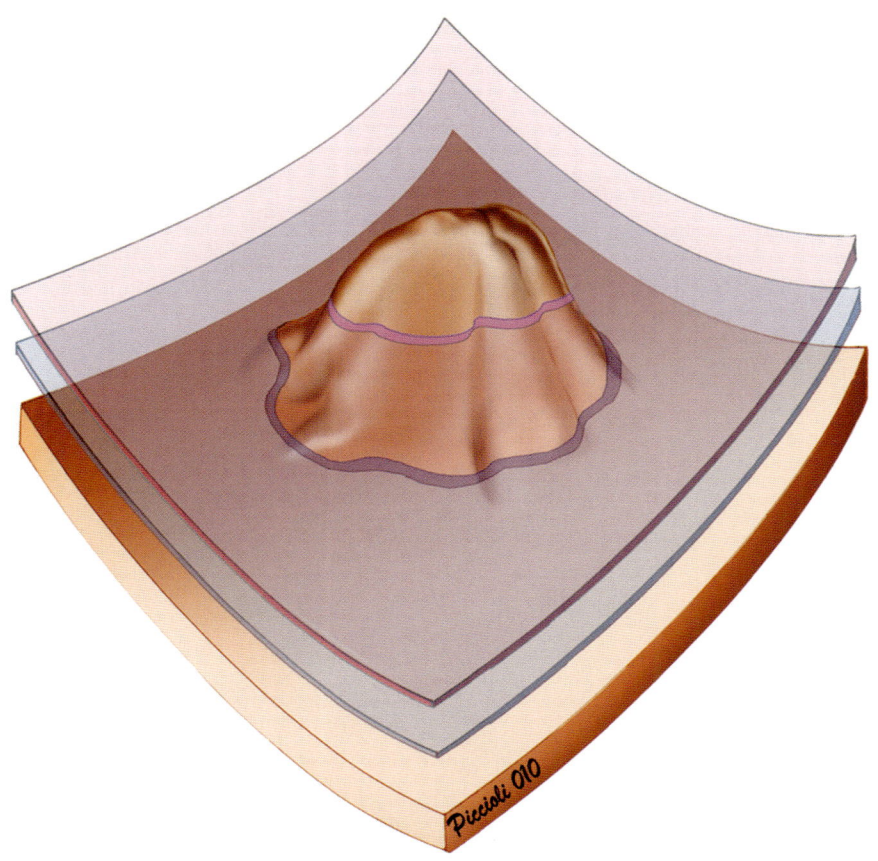

FIGURE 3: FRONTAL, "EN FACE" SECTIONS, FIT TO RETINAL PIGMENT EPITHELIUM CONCAVITY
The sections slice the retinal pigment epithelium elevation, always following the ideal concavity of the fundus.

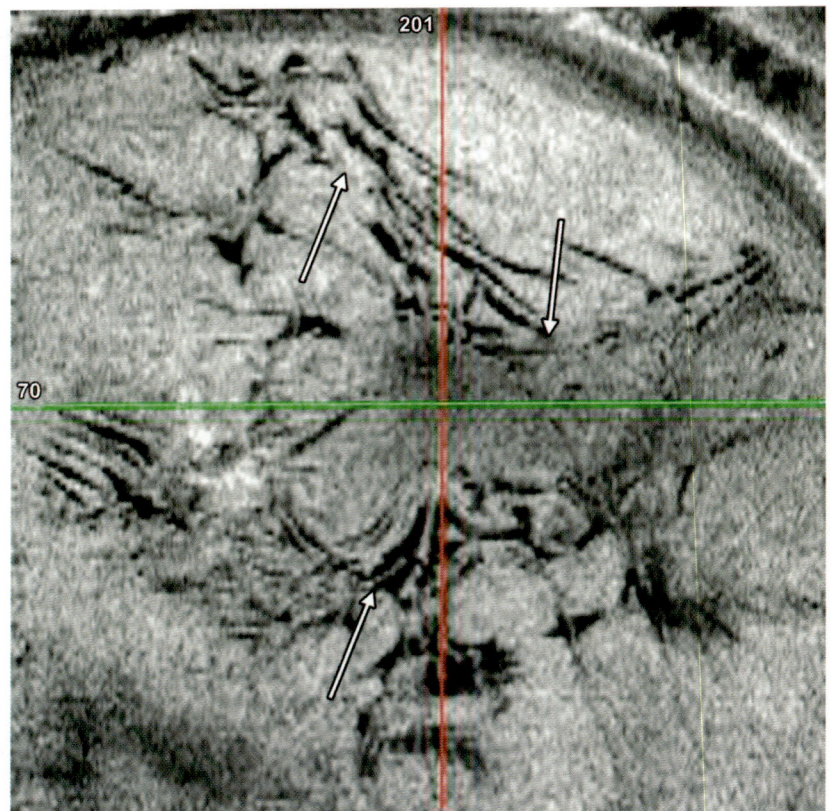

FIGURE 4: FRONTAL, "EN FACE" SECTION, FIT TO INTERNAL LIMITING MEMBRANE CONCAVITY – MACULAR PUCKER

To study the retinal surface we use a frontal, "en face" section, fit to the internal limiting membrane concavity. This section "en face" highlights the retinal folds and their irregular radial appearance. Arrows show the irregular retinal folds.

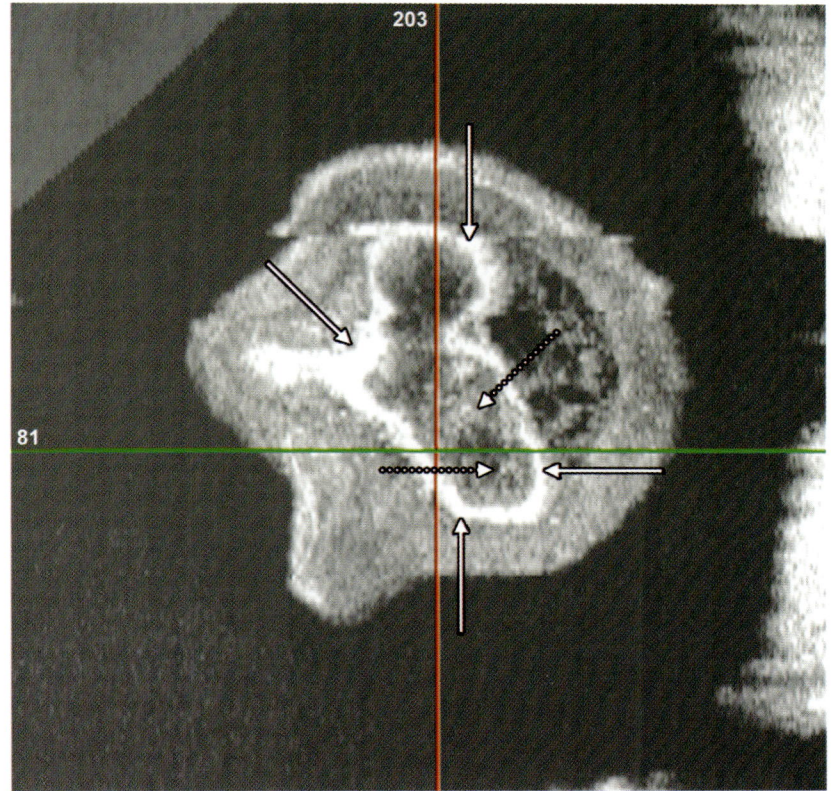

FIGURE 5: FRONTAL, "EN FACE" SECTION, FIT TO RETINAL PIGMENT EPITHELIUM CONCAVITY – VASCULARIZED DETACHMENT OF RETINAL PIGMENT EPITHELIUM IN MACULAR DEGENERATION
The frontal, "en face" section allows us to evaluate the dimensions detachment of the retinal pigment epithelium, the thickness and form of the detachment walls, and the appearance of the detachment contents. The contents show fibrovascular tissue. This section is parallel to the retinal pigment epithelium and slices the choroid 50 μm in front of the epithelium. Line arrows show the pigment epithelium detachment in AMD, dotted arrows show the fibrovascular contents.

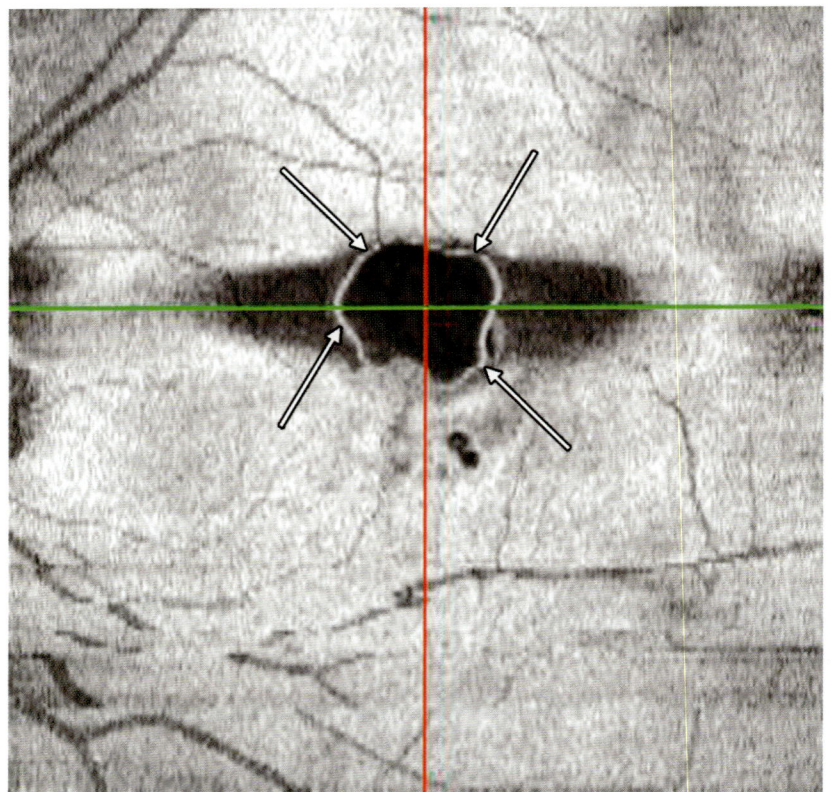

FIGURE 6: FRONTAL, "EN FACE" SECTION, FIT TO RETINAL PIGMENT EPITHELIUM CONCAVITY – POLYPOIDAL CHOROIDAL VASCULOPATHY

This "en face" section highlights a retinal pigment epithelium detachment with walls that are more regular, thinner and smoother than those typical of the vascularized detachment seen in age-related macular degeneration. Line arrows show the pigment epithelium detachment in polypoidal choroidal vasculopathy.

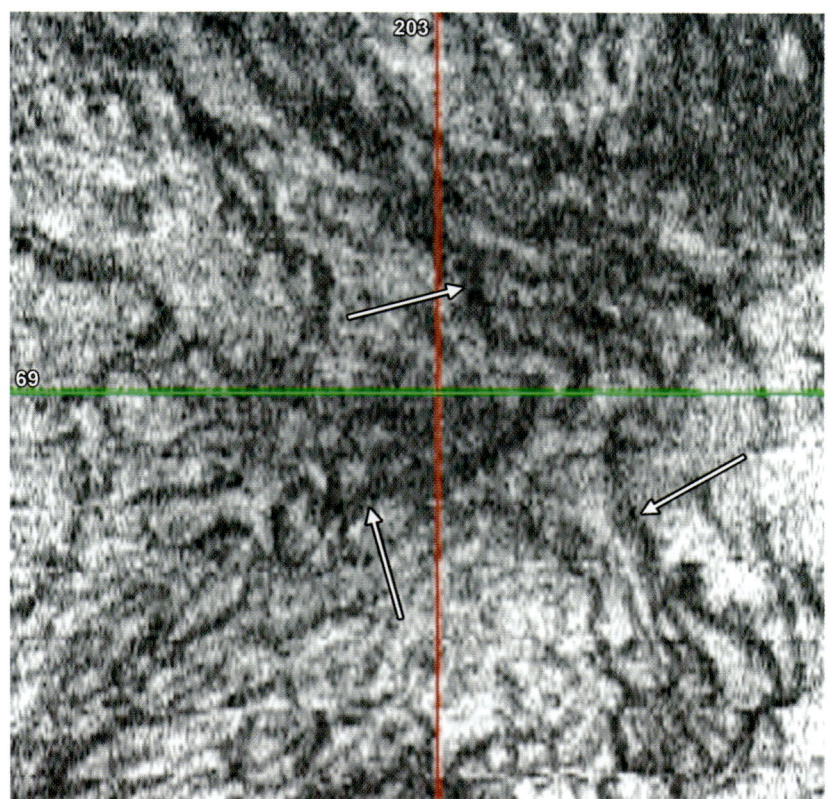

FIGURE 7: FRONTAL, "EN FACE" SECTION, FIT TO RETINAL PIGMENT EPITHELIUM CONCAVITY – HALLER'S LAYER OF THE CHOROID
This frontal, "en face" section, fit to the retinal pigment epithelium concavity, interests only the Haller's layer. Had the section been plane, it would have interested other layers as well. The section is parallel to the retinal pigment epithelium and slices the choroid 90 μm below the epithelium. Line arrows show Haller layer vessels in the choroid.

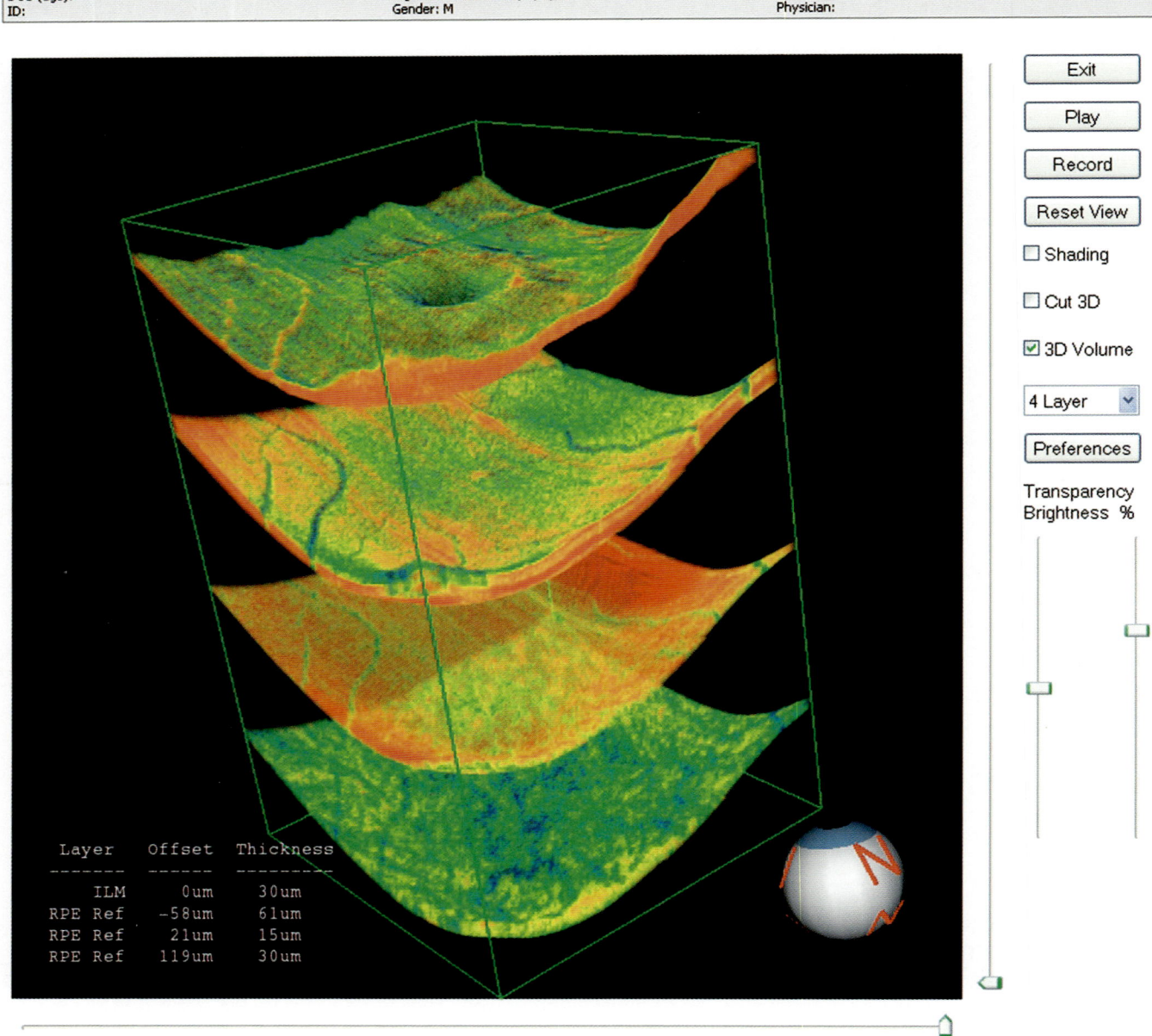

FIGURE 8: FRONTAL, "EN FACE" SECTIONS FIT TO RETINAL PIGMENT EPITHELIUM CONCAVITY – NORMAL EYE

Optovue allows the automatic capture of four frontal, "en face" sections, related to the internal limiting membrane, the retinal pigment epithelium, the choroidal small vessels layer, and the choroidal large vessels layer. These sections can be rotated or slanted in order to study retina and choroid in their full thickness.

Tips and Parameters of the "En Face" Scanning Procedure

- Basic routine allows obtaining high quality "en face" scans.
- **Macula and lesion positioning**.
- **Selection of scan profile (ILM, IPL, RPE or RPEref)**
  - ILM follows the irregularities of the normal and pathologic vitreoretina. It generates 3D images useful to understand the lesion dimensions and its connections with the surrounding structures.
  - IPL follows the irregularities of the normal and pathologic inner plexiform layer.
  - RPE follows the normal profile of the retinal pigment epithelium or the pathologic EPR irregularities; it generates 3D images illustrating the lesion diameter, curvature and dimensions.
  - RPEref eliminates EPR anomalies and provide a theoretical cutting curve that crosses the lesions.
- **Selection of slice thickness (RPEref)**
  - At 10 microns or less, the image is very grainy but highly sensitive.
  - At 16–30 microns, the image is very soft, losing some details in the process.
- **Selection of scanning depth (RPEref)**
  - The fit scan has a constant depth, parallel to EPR.
  - The scans are moved forward to study the retinal layers, 3–5 microns to slice the drüsen, 15–300 microns to cut an EPR detachment, a retinal edema or macular holes.
  - The scans are moved backwards to study the different choroidal layers. The Sattler's layer should be studied with scans at 45–80 microns in depth and slices of 10 microns. The Haller's layer should be studied with scans at 160–190 microns in depth and slices of 20–30 microns.
- Adjustment of contrast and brightness.

CHAPTER 6

Synthetic Study

Within the logic strategy of OCT interpretation, the synthetic study follows the analytical one and constitutes the interpretation most important phase, because only a global evaluation of the various data analyzed can produce a precise and well-founded diagnosis.

An OCT cannot be considered independent from the study of ocular disease. A synthetic evaluation incorporates data that are not directly linked to the OCT.

To correctly interpret an OCT, it is indispensable to know:
- General and local anamnesis
- Patient age
- Cardiovascular examination
- Eye examination complete of vision and biomicroscopic examination
- Fundus photograph with red light
- Fundus photograph with green light
- Fundus photograph with blue light
- Fundus autofluorescence photograph
- Fluorescein angiography. All the sequences of various angiogram times should be studied
- Fluorescein angiography. Manual or automated reconstruction of the retina periphery
- Indocyanine green angiography
- Microperimetry, etc.

Subsequently the OCT analyzed elements should be confronted: morphologic and structural segmentation, high and low reflectivity, as well as:
- Retinal thickness
- Map
- Retinal volumetry.

It also indispensable to:
- Anatomically localize the lesion site in all layers
- Conduct 3D and "en face" examinations.

Thus, we can highlight data pertaining to choroidal or retinal pathology, which would be meaningless in a single frame or isolated scan. In the previous chapters we have separately illustrated the various alterations the OCTs can highlight: these alterations can represent morphologic, structural, reflectivity or quantitative anomalies.

We have to keep in mind that OCT alterations are very irregularly associated.

To these elements we must add the synthetic study of the scans themselves. It is important, and in some cases absolutely indispensable, to use segmentation techniques, tridimensional studies, a retinal map, in order to acquire a complete topographic reconstruction of the fundus.

Diverse elements, once associated, have a diverse meaning and allow considering the ocular diseases in a global fashion (both from the topographic point of view and from the standpoint of the involvement of various eye tissues and membranes).

An isolated study of various tomograms cannot yield precise and global information. The only way to really understand and evaluate OCTs requires a global, omnicomprehensive study of the subject's pathology, avoiding any "guessing", as it unfortunately sometimes happens.

In these last few years, the synthetic study of OCTs has allowed a deeper knowledge of ocular diseases and their physiopathology, helping to discover and highlight pathologic cases poorly known in the past.

> ***A normal OCT can correspond to an abnormal fundus and an apparently normal fundus can correspond to an abnormal OCT.***

> ***The association between qualitative and quantitative OCT alterations is very irregular.***

CHAPTER
7
Reporting

Reporting an OCT Test

As is the case with most instrumental tests in ophthalmology, OCT imaging is not generally diagnostically conclusive if not integrated by other data. In fact, it is indispensable to integrate OCT results with clinical findings (vision, medical examination, tonometry, fundus examination) and instrumental data (FAG, ICG, autofluorescence imaging, ERG, microperimetry), following a logical method of analysis and interpretation while dealing with the large amount of data produced by the tests.

Without clinical data, reporting becomes almost impossible.

The latest generation of instruments offers a number of new views that sometimes make the diagnostic procedure more difficult, instead of simplifying it.

The possibility to generate false negatives limits the performing of OCTs to experienced personnel. The OCT is not a simple photographic picture.

Scans

- **B-scans:** Axial scans, cross-section scans (tomographies).
- Scarce sensitivity, good specificity (it is needed that the scan cuts a lesion in order to analyze it. If the section does not interess exactly the lesion, imaging does not see it. The scans will generate a false negative).
- **C-scans**: "En face" scans, coronal scans (stratigraphies). With devices that support it, it is possible to acquire "en face" scans scans adapted to the curvature of the vitreoretinal interface (ILM), the real retinal pigment epithelium (RPE) and the average curvature ("en face" scan, RPEfit, RPEref). High sensitivity, difficult data to interpret.

The written report of a cross-section scan (qualitative analysis) should begin with the vitreous and proceeds toward the choroid and sclera, crossing all the retina and choroid layers.

For each found element, we should evaluate:
- Morphology
- Structure-segmentation
- Reflectivity
- Abnormal formations.

Subsequently, we should assess the quantitative analysist, basing it on 2D and 3D retinal maps:
- Thickness
- Volume.

High resolution enriches the logical process of B-scan analysis. The retinal tissue is much more detailed and the main retinal layers are visible. The tissue segmentation process takes place.

During the analytic process, we then need to consider the qualitative and quantitative aspects of each retinal "layer", without limiting our attention to the neuroepithelium or the retinal pigment epithelium as a whole.

An OCT report must include:
- At least 2 orthogonal scans that cross the anatomic fovea
- At least one cross-section scan that highlights the pathology characteristics, if the disease does not interest the foveal region
- A macular map
- An "en face" scan analysis of the lesion (planar or fit to the fundus curvature).

If the macular fixation is not good, foveal B-scans will not be centered exactly on the fovea. Vertical scans facilitate the fixation process, because usually the patient can better fix his sight when facing a vertical stimulus (in most cases, microperimetry shows the existence of an alternative fixation area in the sectors immediately above the damaged fovea).

The anatomic fovea is recognizable as the only part of the scan that only visualizes the layer of external nuclei, with a triangular, "apex-like" vitreal representation.

This region can be clearly identified even when the fovea is flattened by a noncystoid macular edema.

The macular map produced with serial (raster) OCT scans is not dependent on foveal centering. Thus, it is much more reliable than the map acquired with scans that are centered on the macula (OCT) and therefore fixation-dependent.

Proceeding from the vitreous to the choroid, we should highlight separately the features of each structure.

The images acquired by Spectral Domain OCT technology require a layer-by-layer retinal analysis.
- **Hyaloid:** Evident, partially or totally detached. Likely evidence of internal limiting membrane.
- **Retinal profile:** Regular, irregular. Likely folds or interruption.
- **Foveolar depression:** Normal or altered in width and depth.
- **Nuclear and plexiform layers**: Evaluation of thickness, reflectivity and possible abnormal structures. Symmetry analysis of adelphous eye.
- **External limiting membrane**: Integrity, discontinuity, abnormal structures.
- **Ellipsoid** (Inner/outer photoreceptor segment junction): Reflectivity, integrity, discontinuity, abnormal structures.
- **Retinal pigment epithelium**: Study of each of the three layers, thickness, integrity, reflectivity, abnormal structures.

- **Bruch's membrane:** (if visible), integrity, discontinuity.
- **Choroid:** Average thickness, vascular caliper, Sattler's layer, Haller's layer, visualization of lamina fusca and superchoroidal space.
- **Sclera:** When visible.

> ***When the time comes to write an OCT report, we advise the young ophthalmologist to refuse to write it in absence of sufficient anamnestic data.***

Bibliography

1. Huang D, Duker JS, Fujimoto JG, Lumbroso B, Schuman JS, Weinreb RN. Imaging the Eye from Front to Back with RTVue Fourier-Domain Optical Coherence Tomography. Slack Inc. 2010.
2. Lumbroso B, Huang D, Rispoli M, Romano A, Coscas G. Clinical "en face" OCT Atlas, Jaypee Publisher, 2013.
3. Lumbroso B, Rispoli M. Practical Handbook of OCT, Jaypee Publisher, 2014.

CHAPTER
8
Glaucoma

Glaucoma is an optic neuropathy that involves ganglion cells loss and damage to the nerve fibers layer and is frequently associated to elevated eye pressure. The main role of OCT for glaucoma study, is to investigate retinal nerve fiber layer (RNFL) and ganglion cells complex thickness. In this chapter we use RTVue Optovue Inc., Fremont, CA, but there are no significant differences among other modern OCT instruments for detecting RNFL loss.

Glaucoma Scan Protocols

To study glaucoma RTVue uses four scan modalities.
1. **Optic disk scan protocol**: The disk is calculated from a 6 mm × 6 mm cube, formed by 101 lines. 3D disk scan is used as baseline and to identify rim and disk vessels. 3D baseline is the disk boundary that is automatically drawn by the software using the 3D disk scan.
2. **Retinal nerve fiber layer scan:** This is used to study nerve fiber thickness. The RNFL scan pattern completes four circular scans around the optic nerve head. These scans are averaged and the result is presented within the normative range parameters.

    Retinal nerve fiber layer thickness profile in the ONH is the thickness of RNFL at a calculated 3.45 mm diameter around the center of the disk. Eventually the de-centering of the disk will not affect the measurement. The RNFL thickness map is color coded where thicker RNFL values are brighter colors and thinner RNFL values are darker colors.

    The TSNIT graph, plotting the RNFL thickness profile going around the optic disk starting temporally and moving superiorly, nasally, inferiorly, and back to temporal again. This thickness profile is shown as the black line superimposed on the normative database values, with normal being the green area and outside normal the red area.
3. **Optic nerve head map:** This protocol completes 13 circular scans around the optic nerve head, and 9 radial scans. They delimit the space between pigment epithelium and optic nerve head. The device calculates automatically the center of the disk. Optic nerve head map gives important disk morphology information: Disk and Cup Areas, Cup/Disk Ratio, RNFL 3.45 and NFL thickness map from disk margin up to 2 mm radius from the center of disk.
4. **The GCC (Ganglion cell complex) map:** The GCC scan provides important parameters for glaucoma diagnosis. *Nerve fibers layer, ganglion cells layer and inner plexiform layer form the*

ganglion cell complex in the macular region. The diagnosis of glaucoma is improved when we study the ganglion cell complex rather than the entire retinal thickness. Ganglion cell complex study is useful for early diagnosis of glaucoma and to assess disease progression. Ganglion cell loss occurs before visual field lesions and before nerve fiber layer thinning.

Ganglion cell complex thickness is defined by the distance from internal limiting membrane (ILM) to outer inner plexiform layer (IPL), which composes the inner 3 layers of the retina [NFL, ganglion cell layer (GCL) and inner plexiform layer]. As the ganglion cell dies, the GCL becomes thinner.

It can display the thickness map and deviation from normal of the inner retina.

The **deviation map** shows the deviation from normality for age and ethnicity as presented by normative database.

The **significance map** reports the patient findings compared to normal ethnicity as presented by normative database. It uses red yellow and green color to note the changes during the follow up. It indicates the degree of significance, that is: how important is the difference with normality.

Artifacts: Eye motion and blinking can cause artifacts. In the maps the eye movement appears as a break in the blood vessel pattern. If such breaks are seen, the scan should be repeated.

Glaucoma Follow-up

Progression analysis compares RNFL thickness measurements and ganglion cell complex maps over time and determines if statistically significant change has occurred. The rate of change is quantified.

Ganglion Cell Complex Glaucoma Progression Report

The GCC progression report can display up to 4 scans at a time (1 baseline and 3 follow-up scans) separately for each eye. The printout can be divided into 4 sections.
- The first 3 sections provide the GCC thickness map (color-coded thickness map representing the thickness of the GCC of the macular region).
- The deviation map (color-coded map showing the percent loss from normal as determined by the instrument normative database).
- The significance map (color-coded map showing regions where the change from normal reaches statistical significance, with green, yellow, and red representing values within the normal range, borderline and outside normal limits to allow visual comparison across visits).
- The fourth section contains a "fit each value" graph for the average, superior and inferior GCC (i.e. the GCC thickness values for the full retina and for the superior and inferior hemiretinas, respectively, with the exclusion of the circle mask in the fovea). The section also shows a table of the result for baseline scan and each follow-up scan, along with the change in microns calculated between the baseline and the last follow-up scan.

Retinal Nerve Fiber Layer Glaucoma Progression Report

The retinal NFL 3.45 protocol assesses the peripapillary retinal NLF thickness by placing a 3,45 circle scan around the optic. The retinal NFL 3.45 is the average of 4 circle scans.

The retinal NFL progression printout can be divided into 3 sections. In the first section, sectoral retinal NFL thickness result and SSI values at baseline and each subsequent visit are reported. In the second section, the TSNIT graph is displayed for visual comparison of the retinal NFL double hump pattern across scans. The third section contains a fit each value graph for the average, superior and inferior retinal NFL thickness and a table showing the result for baseline scan and each follow-up scan along with the change in microns calculated between the baseline and the last follow-up scan.

Normative Database

Normative database makes it possible to discriminate normal conditions from pathological ones. The OCT devices differ in the number of eyes included in. They include data for both retina (edema and ischemic effects in the macula) and glaucoma (Optic disk, ppRNFL, NFL to 4 mm and ganglion cell complex). They help comparing patient measurements with age-matched normal subjects. The RNFL normative database uses color to indicate the normal distribution among individuals of the same age. Red: indicate that readings are pathological, outside normal limits. Yellow: indicates borderline readings, they fall inside the yellow zone and are considered suspects. Green measurements are normal.

The normative database is used to provide a relative comparison of where a particular patient's results fall within the parameters of the "normal" population range for their age and ethnic group. The color coding for the normative display uses a green (within normal range), yellow (borderline range) and red (outside normal range).

It is important to know that normative database comparisons are based on statistics only and that it is possible to find normal persons outside normal database range.

Glaucoma Diagnosis

Figures 1 to 4 show the particular features in different glaucoma stages analyzed with Optovue OCT.

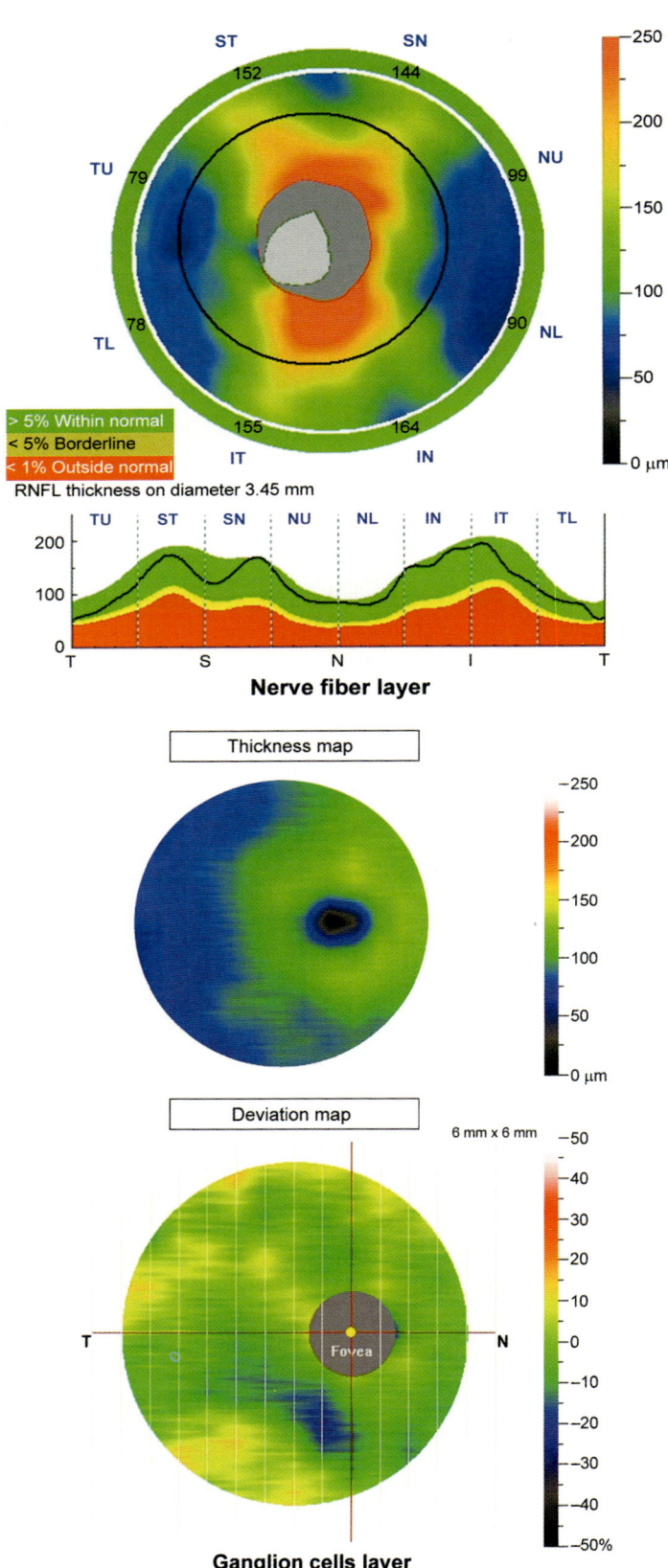

FIGURE 1: EARLY GLAUCOMA

GLAUCOMA

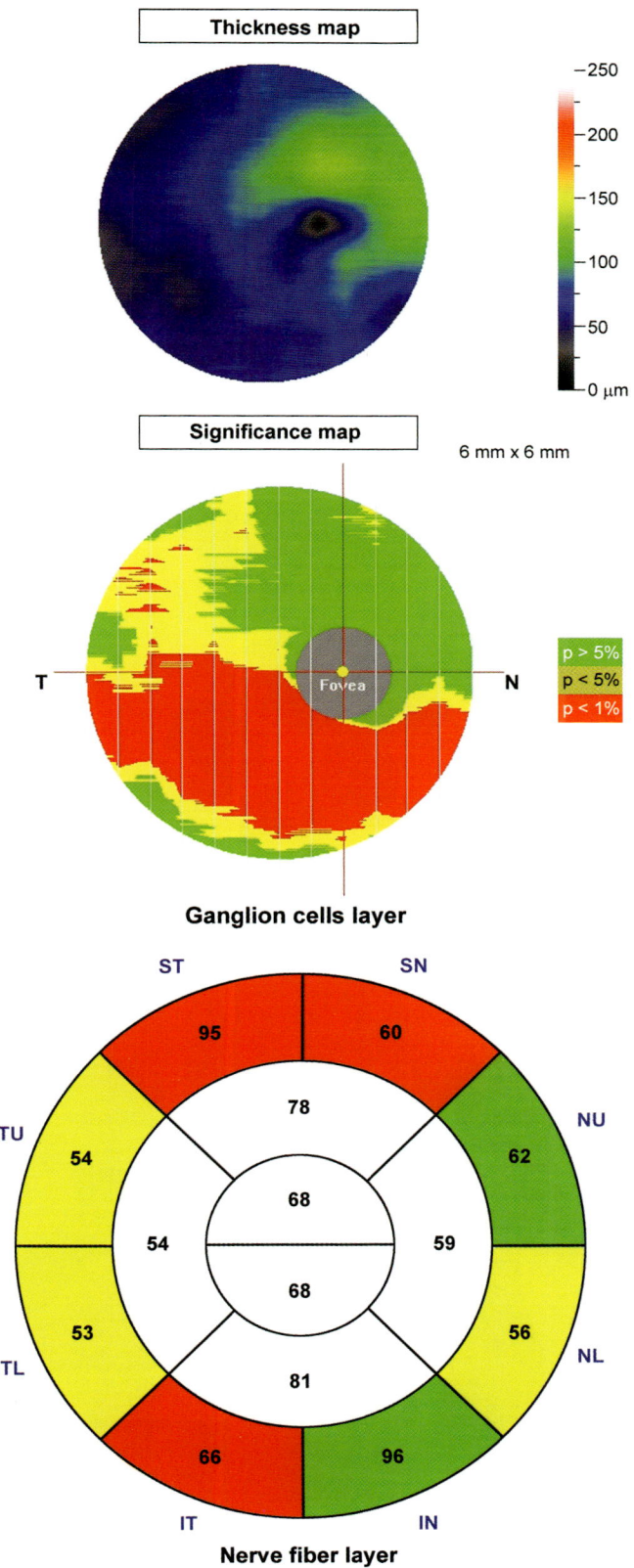

FIGURE 2: MODERATE GLAUCOMA

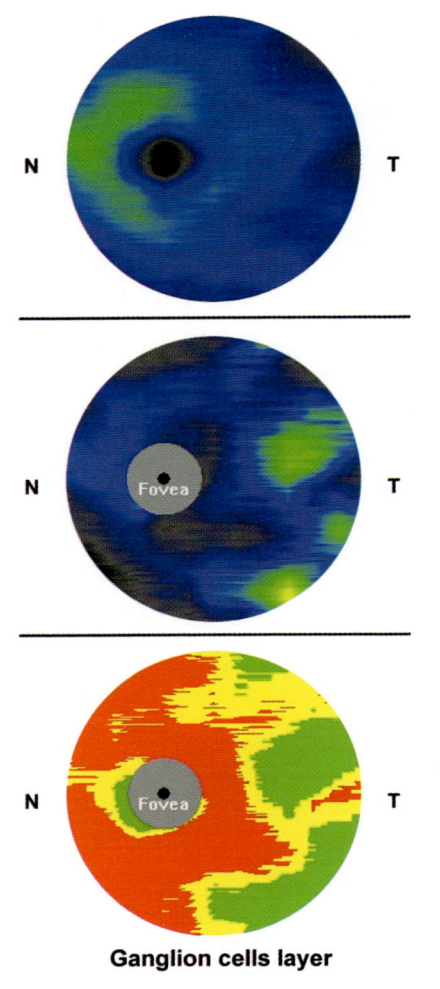

FIGURE 3: SEVERE GLAUCOMA

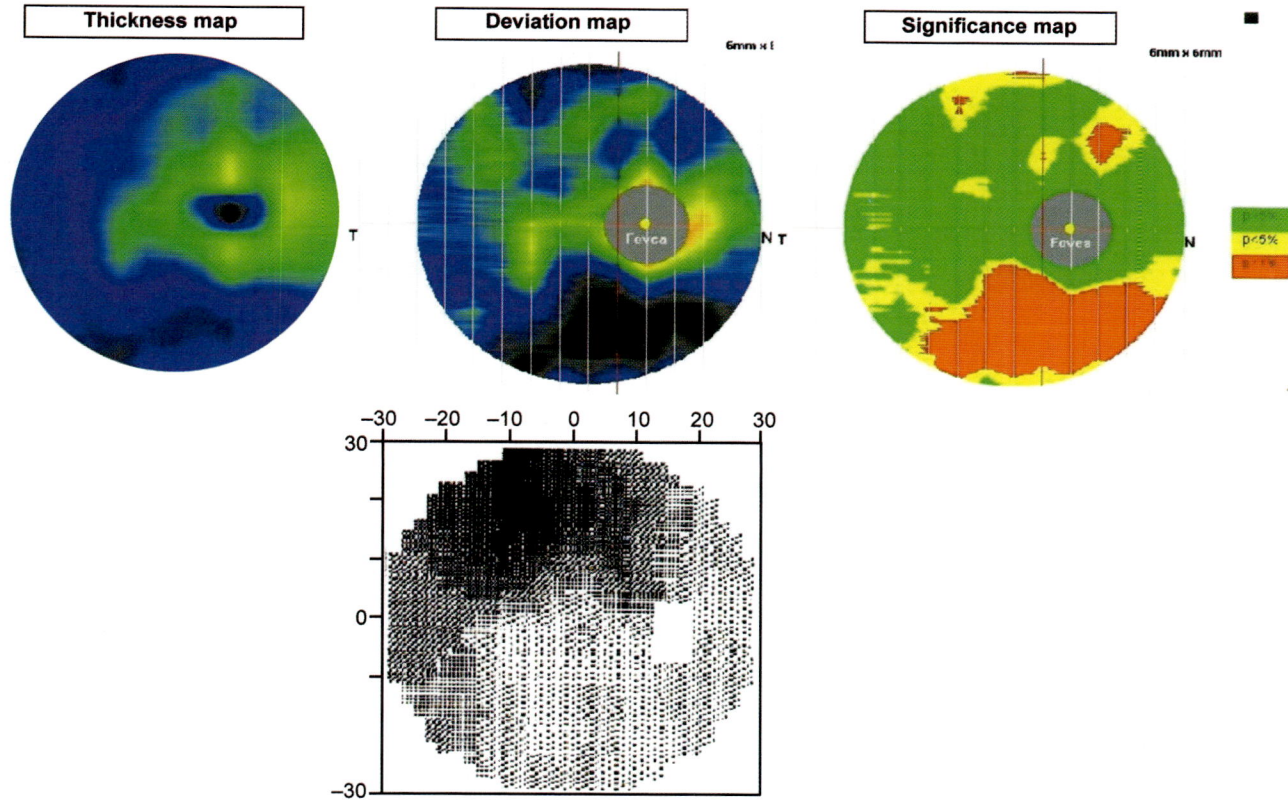

FIGURE 4: GANGLION CELL COMPLEX THICKNESS MAP, DEVIATION MAP, SIGNIFICANCE MAP AND VISUAL FIELD IN A GLAUCOMATOUS PATIENT
The visual field alterations correspond to the ganglion cell complex alterations in the RTVue scan.

Index

Page numbers followed by *f* refer to figure and *t* refer to table

A

Acute zonal occult outer retinopathy (AZOOR) 42
Adherent epiretinal membranes, morphologic alteration 35
Alterations, morphologic 39*t*
Analytic qualitative of high reflectivity abberation 58
Artifacts 19
Asymmetric lamellar hole, morphologic alterations 38

B

Backscattering 49
Bipolar
 cell 16
  nuclei 42
Bruch's membrane 3, 24, 30, 45*t*, 84

C

Cell chains 15, 41
Cellular chains 3*f*
Chorioretinal OCT analysis, normal 11
Choroid 3, 19*f*, 39, 59, 84
 normal 40*f*
Choroidal 80,
 layer 5*f*
 nevus 60
 reduced thickness, causes of 18
 thickness variations 66*t*
 vessels 18
Classical neovascular membrane, abnormal 57
Clinical interest of 3D OCTs 68*t*
Concavity 22
Cone, normal shadow 59
Convexity 23
 caused by diffuse retinal edema 26*f*

Coronal images, measurement of 61
Cotton wool exudates 55
 causes of 57*t*
Cystic space contents 51*t*
Cystoid
 edema 50
 macular edema, causes of 50*t*
Cysts 50

D

Deep morphologic alterations, high reflectivity 39*f*
Deep retina level 59
Diabetic maculopathies 50*t*
Diffuse edema in diabetic retinopathy 26*f*

E

Ellipsoid 45, 83
 inner 47*f*
 lesions of 48*f*
 outer 47*f*
 zone 10
En face
 imaging 67
 section 48*f*
Epiretinal membrane *See* Preretinal membrane
Epiretinal membrane, morphologic alterations 36
External limiting membrane 10, 20, 42*t*, 83
Exudates – abnormal formation 56*f*
Eye
 membranes 16, 81
 tissues 81

F

Fluorescein angiography 80
Frontal "en face" sections fit to RPE concavity 68, 75 to 78*f*

Frontal plane scans 67
Frontal, "en face" section, fit to internal limiting membrane concavity 74
Fundus, reconstruction of 81

G

Ganglion cell 15
 complex 1, 85-87
  glaucoma progression report 86
 segmentation 41
Gass classification 23
Glaucoma 85
 diagnosis 87
 follow-up 86
 scan protocols 85
Global retinal deformation 22

H

Haller's layer of large vessels 18
Hard exudates, causes of 57*t*
Henle fibers 2
 radial structure 6*f*
Hyaloid 83
 membrane 1

I

Inner layers, lesions of 44*t*
Inner retina *See* Ganglion cell complex
Intraretinal formations 55
 hard exudates 55
 hemorrhages 55

L

Lamellar hole 23
Layers and photoreceptors, lesions of outer 44*t*
Lesion positioning 79

Linear measurements of axial
B-scans 61

M

Macula 79
Macular
   cystoid edema 33
     advanced 34
     differential diagnosis of 50t
     in diabetic retinopathy 33
     structural alterations 52
   hole–morphologic alteration 36
   holes 23
   pseudohole 23
   pucker 55
     morphologic alterations, retinal 35
Membranes
   external limiting 16
   internal limiting 16
Microcysts 50
Morphologic posterior alterations 24
Müller's cell 4f, 9f, 15, 16, 41
   fibrillae 16
Myopia 22
Myopic retinoschisis 34

N

Neovascular membranes, causes of 59t
Neuroepithelium detachment
   low 34
   of chronic epitheliopathy 52f
Normative database 87
Nuclear layer 42t, 83
   inner 1, 42t
   outer 10, 42t

O

OCT
   and histology 1
   limits 22
   normal 11
   of optic disk 22
   of retinal structure in gray scale 13
   terminology, new 10f
   test 82
Ocular disease 80
Optic
   disk scan protocol 85
   nerve head map 85

P

Photoreceptor segment junction
   inner 42
   outer 42

Pigment
   deposits–abnormal formation 58f
   epithelium collection detachment 50
Pigmented cell 2
Plexiform layer 15, 42t, 83
   inner 1, 16
   outer 2, 10, 16
Preretinal membrane 55

Q

Qualitative analysis 23t
   of pathologic OCTs 21
Quantify fluid collections 49
Quantitative
   analysis of pathologic OCTs 61
   analysis: thickness 63
   analytical retinal study 66t
   choroid analysis 66
   optic disk analysis 66
   segmentation 62

R

Reflectivities
   fluid collection, low 49
   high 48
   in inner and outer retina, high 48
   low 18, 49
   of normal tissues 18t
   preretinal formations, low 55
   structures, low 51t
   study 48
Retina
   and choroid
     morphology 11
     structure 11
   division between outer and inner 16
   inner 1, 41
   outer 1, 41
   outline, outer 30
   periphery, reconstruction of 80
   surface deformation 23
Retinal
   architecture 18t
   atrophy–deep high reflectivity 49
   barriers, horizontal 49
   cells form chains of photoreceptors 3f
   folds caused by horizontal traction 27f

histologic sections 16
layers 5f, 9f
   horizontal structures 16
   thickness variations of 65
mapping 62
nerve fiber layer
   glaucoma progress 87
   scan 85
pathology See Choroidal
pigment epithelium 2, 20, 19f, 24, 32, 45, 49, 59, 82, 83
   and inner plexiform, layers between 42
   detachment 54f
     of occult neovascular membrane in age 54f
   drüsen 31
   drüsenoid detachment of 31
   tear 32
profile 23, 24, 83
   cross-section scan 38t
   deformed by vitreoretinal traction 28
section, thickness of 68
structure study, segmentation 41
support structures 2
topography–occlusion of arterial branch 71f
volume 65
Retinoschisis, causes of 43t

S

Scan profile, selection of 79
Scanning depth, selection of 79
Scar 59
Sclera 19f, 84
Segmentation 16f, 19f
   of chorioretinal layers 67
Serous
   detachment of neuroepithelium 6f, 30
     with elevated photoreceptor alterations 7f
   neuroretinal elevation, causes of 52t
   retinal pigment epithelium detachment, causes of 55
Shadow
   areas—screen effect 59
   effect—screen effect 60t

INDEX

Slice thickness, selection of 79
Subatrophic choroid 40f
Subretinal
   collection 50
   deposits 44t
   layers 23
Synthetic study 80

T
Tamoxiphen poisoning 42
Tissue reflectivity, normal
   choroid 18
   high 18
   low 18

V
Vertical structures 15

Vitreoretinal
   anteroposterior traction 37
   interface, curvature of 82
Vitreous 1

W
White dots syndrome, multiple evanescent (MEWDS) 42